蔓·设计

Organically-Permeated Design

王振军 著

中国建筑工业出版社

图书在版编目（CIP）数据

蔓·设计 / 王振军著. 北京：中国建筑工业出版社，2017.11
ISBN 978-7-112-21379-5

Ⅰ.①蔓… Ⅱ.①王… Ⅲ.①建筑设计-文集 Ⅳ.①TU2-53

中国版本图书馆 CIP 数据核字 (2017) 第 252786 号

责任编辑：毋婷娴　张伯熙
责任校对：王　瑞　焦　乐
策划编辑：方　雪

蔓·设计
王振军　著
*
中国建筑工业出版社出版、发行（北京海淀三里河路9号）
各地新华书店、建筑书店经销
中国电子工程设计院王振军工作室制版
北京顺诚彩色印刷有限公司印刷
*
开本：965×1270毫米　1/16　印张：13¼　字数：318千字
2017年10月第一版　2017年10月第一次印刷
定价：288.00元
ISBN 978-7-112-21379-5
（31104）

版权所有　翻印必究
如有印装质量问题，可寄本社退换
（邮政编码　100037）

王振军工作室主持人 王振军

目录
CONTENTS

蔓设计 / Organically-Permeated Design　06

作品 / Works

1. 2010 年上海世博会沙特国家馆　15
   The Saudi Arabia Pavilion at Shanghai EXPO 2010

2. 上海浦东软件园一、二、三期　35
   Pudong Software Park, Phase I,II,III in Shanghai

3. 中国信达（合肥）灾备及后援基地　51
   China Cinda (Hefei) Disaster Recovery and Back-up Base

4. 长沙中电软件园总部大楼　65
   The Headquarters Building of Changsha CEC Software Park

5. 海盐杭州湾智能制造创新中心　79
   Haiyan Intelligent Manufacturing Innovation Center

6. 北京新机场系列项目　93
   Beijing New Airport Series Projects

   (1) 北京新机场信息中心和指挥中心　95
   The ITC and AOC for Beijing New Airport

   (2) 北京新机场行政综合业务用房工程　103
   Beijing New Airport Administrative Integrated Business Building Project

   (3) 北京新机场生活服务设施工程　109
   Beijing New Airport Living Service Facilities Project

   (4) 非主基地航空公司办公及宿舍项目　115
   Non-main-base Airline Office & Dormitory Project

7. 首都国际机场东区塔台　119
   Eastern Air Traffic Control Tower of Beijing Capital International Airport

8. 西安咸阳国际机场新塔台及附属建筑　123
   The New Air Traffic Control Tower of Xianyang International Airport and Its Affiliated Building in Xi'an

9. 郑州新郑国际机场新塔台及附属建筑　129
   The New Air Traffic Control Tower of Xinzheng International Airport and Its Affiliated Buildings in Zhengzhou

10. 北京按摩医院　139
    Beijing Massage Hospital

11. 阿克苏三馆及市民服务中心　147
    Aksu Public Complex and Civic Service Center

12. 巴黎世界环保展示中心及企业孵化器　155
    The World Environmental Protection Exhibition Center and Business Incubator in Paris

13. 郑州航空港银河办事处第二邻里中心小学　161
    Zhengzhou Airport Yinhe Office No.2 Neighborhood Central Primary School

14. 北京雁栖湖山地艺术家工作区　169
    Beijing Artist Studios on Yanqi Lakeshore

15. 美国塞班岛马里亚纳别墅区　175
    The Mariana Villa Area in Saipan Island USA

蔓设计再思考 / Rethinking Organically-Permeated Design

人与自然通过科技在信息时代的整合 —— 上海浦东软件园设计　　180
Integrating Human and Nature via Science and Technology in the Information Era —— The Design of Shanghai Pudong Software Park

探索软件园设计的生态学途径 —— 上海国家软件出口基地规划设计　　184
An Exploration into Ecological Approach for Software Park Design —— Planning and Design of Shanghai National Software Export Base

回归本质　　190
Back to the Essentials

附录 / Appendix

作品年表 Chronology of Works	194	工作室的蔓设计 The Organically-Permeated Design by WZJ Studio	204
作品获奖 Awards	200	工作室简介 Introduction of WZJ Studio	208
媒体评价 Reviews	201	工作室成员 Members of WZJ Studio	208
参展 Exhibitions	202	致谢 Acknowledgements	209
论著 Treatises	202		

蔓设计

文 / 王振军

关键词：蔓；蔓延；蔓设计；有机；有机物态与有机建构；自然场力；浅层次与深层次的建构；拓扑心理学；有机建筑与新有机建筑；建筑的可持续性

一、概念提出的背景及其涵义

1.1 问题的提出——理念与方法

对于大多数建筑师来说，职业生活就是这样，眼看着整天忙忙碌碌，人就像在生产线上一样，图纸在一套套地发出去，城市却一天天地在平庸中沉沦，而你被无尽的、各式各样的纷扰包围着，什么也做不了，只能拿着心中偶像的作品集发愣，或者等待着一场逃离。

2009年成立工作室以来，我就一直在设想从此摆脱上面所说的状态，不再像往常做常规项目那样，而可以进入一种研究、思考而后实践的状态，全情投入、从容不迫地去完成自己喜欢的项目，不太容易被紧迫的时间和其他因素所打断，从而能把对项目的哲学思考、建筑学意义上的追求完全贯彻到项目的全过程，实现一个完成度相对较高、较少存在遗憾的作品。由此，我开始琢磨是否有一个更凝练的说法用以秉承，同时能够把这种诉求清晰地表达给团队甚至甲方。

带着这个问题，在一次工作室的务虚会上大家进行了一番讨论。首先就有同事发言，"好的设计一定是在放松的、深入的状态中循序渐进的，但又不是简单的慢慢设计，在目前的中国设计生态下，原本也不允许建筑师慢慢地做设计"。"慢"字进入了讨论的话题中，的确，我们不可能慢慢地设计，但是需要让理念蔓延或者说延伸到始终的设计。经过20年高速发展城市建设，"快餐式"的设计出来的建筑充斥着城市各个角落，城市需要我们拿出更高完成度的作品、客户也希望我们拿出较以往更加精致的成果，并且我们内心也渴望精雕细琢出无愧时代的作品，以期能在未来更高水准、更严酷的市场竞争中赢得尊重。

提到màn延中的"màn"字，在新华字典中有曼、漫和蔓三种说法，分别意味着直线、平面和立体三种扩展状态，毫无疑问，立体化的"蔓"更能贴切地表达建筑师的想法。

"蔓"呈现的有机物态和生命力恰如其分地与工作室一直以来所秉持的、以赖特为代表的有机建筑理论为发端而又不断被丰富的有机建构理念相契合。

由此看来，"蔓设计"的确很贴切地表达了工作室对建筑创作在理念和态度上的诉求。

1.2 蔓设计的涵义（全书文中提及的"蔓设计"同书名"蔓·设计"）

1.2.1 蔓 —— 有机物态与有机建构 —— 一种理解自然、看待自然、介入自然的态度和方式，一种设计哲学。

这的确不是一个新概念，而是一个从赖特以来被不断丰富的理论。当我们俯瞰每个项目的基地，或仔细观察其中的一树、一叶、一石的纹理，不难发现其中充满了自然力的和谐统一。世上的一切技术和艺术形成皆可从自然界的各种生物存在形式中推演出来。但是人类却总自诩为世界上最完美的结构，总苛求世界上的一切事物都完美起来，尤其在今天，人类的能力已无法控制地膨胀起来，使这种追求完美的欲望变得更为强烈。

建筑作为人类活动的一种产物，充分地反映了人类的这种活动规律，城市和建筑这种人工物态正是人类在不断追求和探索完美中发展起来的。不幸的是建筑和城市在显示人类力量和伟大的同时，当下结果却并不令人满意，建筑这一人工物态和自然环境越来越呈现出矛盾和对抗，而且这种对抗还在呈现出加剧的状况。自然创造了人类，人类又创造了建筑和现代化的城市以及现代生活。但不幸的是，现代生活使我们越来越远离了自然并招致对自然的无数曲解，其结果是使我们失去机会来把握这一能量的本质。

法国科学家克洛德·阿莱格尔（Claude Allègre）在《城市生态，乡村生态》一书中写道："……现在的人类必须明白，开采的时代结束了，展现出来的是一个管理与保护的时代，对抗的时代过去了，

图1：自然场力

展现出来的是和睦的时代。"

（1）"蔓设计"的有机建构理念

毫无疑问，人类又到了需要认真审视建筑与自然关系的时候。建筑哲学上出现了"回归自然"的热潮，但有两个倾向值得我们警惕，一是有的建筑师随意套用钱学森的"山水城市"，无论什么规模建筑都直接处理成山水样子的做法，二是肆意冠以"绿色建筑""建筑节能"的名号。

"蔓设计"的有机建构理念崇尚立体化地去实践我们与自然的这种相处模式而不是表象化的模仿（也许一个方盒子在某地某时是有机的）和机械化填充，因为大自然本身就是一个极为复杂而有机的系统，出于成本和情感的原因建筑师应继续发挥人类简化和抽象的本能。因此基于有机建构理念的"蔓设计"就应是一个面向未来的、开放的、不断丰富的理论系统。

（2）自然的有机物态和人工的有机物态的共同特点——多元、有机、深层次、系统性

①有机的涵义：首先是指形成一个复杂的实体，该实体的整体比各个单独的部分更为丰富，而各部分也具有因其参加了整体才有生命的特征。其特点是整体性和生命性。

②有机物态关系的定义：物态之间应该是各具个性，而又共存于某种因素的联系之下。如自然物态在各种场力（包括阳光、风、雨、雪、热、冻、地球引力、地震力等力）的作用下所呈现的有机状态，也指一种有机物态自身发展的规律（生、老、死、再生）（图1）。

③人工物态（建筑）作为一个后来者，一旦介入自然界就应该与自然界已有的有机物态发生关系，这种关系处理的质量取决于建筑师，需要建筑师借助自己的知识、灵感和智慧去进行，从而与既有物态保持一个有机物态关系而不是生硬或疏远的关系。

（3）建筑的可持续性概念的厘清

①这不是一个口号，它主要强调思想和技术，是一种全面的建筑观，不是一个建筑流派思想，不纯粹强调形象和风格。

②既要考虑当下又要兼顾未来的项目全生命周期。

③它是指项目建设完成的全过程，而不仅仅指设计阶段。

④始终要思考怎样借助自然的恩惠，对一切自然资源加以充分利用。

1.2.2 蔓设计方式——一种持续地将设计理念贯穿项目始终的工作方式

（1）蔓设计，不是慢设计，是蔓延始终的设计——设计理念贯彻始终的思考、工作方式和手段。

（2）蔓延全过程：从策划、规划、单体方案、初步设计、室内设计、景观设计、施工图绘制、施工监控等全过程贯彻。

（3）蔓延全手段：从二维到三维、从平面化图纸到立体化模型……全方式的推敲与优化。

二、分析与思考

2.1 心理因素与有机物态关系

2.1.1 拓扑心理学（Topological Psychology）的启示

提到人与环境的关系问题，必须应涉及拓扑心理学创建人德国心理学家勒温（(Kurt Lewm）提出的公式 B=f(PE)，认为行为（B）等于人（P）和环境变化系数（E）的函数，有些事物吸引人，具有引值（正的原子值），是人所愿意接近和取得的；有些事物排拒人，具有拒值（负的原子值），是人所不愿意接受或拒绝的。

2.1.2 如何从建筑建构关系的基本构成特点出发来满足人的心理需求，以及心理因素对物态关系反映的复杂性与不定性的诉求，是我们在创作中要思考的问题。局部应尽可能强调其应有的特性，而整体上强调

图2：美国国家大气研究中心（建筑师：贝聿铭）
（图片来源：美国国家大气研究中心官网）

相互间组合关系和排列方式，也就是我们这里所说的建筑建构关系上的有机物态关系。

2.2 对现在与将来要解决的问题的探讨

2.2.1 现代主义方盒子的局限性及其解体模式

人类对自然物态简化抽象的本能和技术的进步使这种能力达到了极限，而现在需要回归去追求有机物态。

（1）解体模式产生的必然性

由"现代建筑原则"而导致的千篇一律及有机物态带给我们的启示，建筑的确具有开始多元化的必要性，科技和思维的发达更是为这种多元化带来保证。各种原则带来统一的同时也与科技和思维的发达带来的多元化相矛盾、与丰富的自然界相抵触，人类渴求自身真正永远置身于流变的、美好的有机物态关系之中。

（2）从有机物态关系考察所想到的问题

有机物态关系产生于自然场力、人类改造环境的能力及心理因素产生的场力的高层次、多方位的结合。据此，建筑设计应是，在寻找维系因素（各种场力）并思考未来发展变化的前提下强调个性。我们的创造只有维系在这种场力中才会有活力。贝聿铭先生当年设计美国国家大气研究中心时的做法就是一个范例（图2）。为了了解基地，他曾携带睡袋在基地过夜并开车在沿途附近的印第安岩石构造遗址考察，从而为设计汲取灵感。西扎（Alvaro Siza）的博阿诺瓦餐厅（图3）通过对场所特有的地貌和微型地理学的形式构成进行提炼，建筑的平面、体量和屋顶形式源于对布满岩石的海角场地的细致研究。他的建筑能让人强烈地感受到是从其场所中抽取出的各种向量的自然场力集合并相互作用的结果。

（3）对传统的地方建筑和对自然简单化模仿倾向的批判

由于科技水平落后，传统的地方建筑是完全屈从于自然场力的一种被动消极的适应，需要在其中注入当代科技的能量加以改进，例如现在有些建筑师搬来几十年前的生土做法来充当设计的噱头，这种既浪费人力又损耗土地资源的做法令人错愕。而形式化地对自然物态的模仿更是对有机物态建筑的误读。这种"强人所思"的做法是一种"机械式的""浅层次的"创作态度。

2.2.2 对中国建筑发展的思考

（1）当今及未来建筑的建构关系应是主动、积极的有机物态关系的建构。当今先进的技术和高度发展的哲学使我们有能力去剖析各种有机关系的产生、变化和发展，并积极思考怎样使其交融于建筑。

（2）对转换"场力因素"的思考

利用不同基地"场力因素"的变换，创造积极、主动的、丰富的、具有有机物态关系的建筑。思考原则虽然一致，但由于基地的千姿百态，不必担心设计成果会趋同。

（3）深层次有机物态关系的思考

今天的人们已经具有透过表象观察事物的内在关系及规律的能力，甚至达到从"客体"到"精神"的超脱，而人类的这种能力展现在建筑创作上，就是对建筑建构关系深层次的思考，也就是在追求产生深层次的"物态效应"的场力因素影响下的有机物态关系。

2.2.3 对中国建筑设计的议论

（1）中国当下的建筑乱象。建筑作品的首要服务对象不再是客户而是公众和社交媒体，每一个试图在舆论中生存的建筑师将受到考验，建筑从未像今天受到关注，可谁也不知道这意味着什么？"……建筑师只是设计风格的印钞机……建筑学的评价体系变得诡异而且不可捉摸，建筑师既可是巫师也是道士，建筑师的理念只是房地产的炼

图3：博阿诺瓦餐厅（建筑师：西扎）

金术","当众多建筑理论家相信建筑形式的不确定性是新建筑的潜在因素时，同样多的时尚元素也将在建筑中使用来诠释时代的意义"。

（2）只有中国人才能创造中国人自己的建筑，我们自己应立足当代去造就具有现代感、对当代中国社会变化做出回应的中国建筑。目前，中国当代创作的一些建筑作品确实呈现出"以他人之新为新"、"鹦鹉学舌"的现象，技巧和手法固然重要，但由于缺乏理念和原则，长此以往会使自己永远停留在技法的层面，而最终迷失在西方的评判标准中。造就当代中国建筑的责任还是在中国建筑师的肩上。

（3）建筑学的真正创新还未到来，因为还没有一种普遍知识主宰我们。但建筑学的基石至今未动，我们仅处在语言创新和手段创新的时代。地方（域）主义解决不了普遍问题，所以是虚假的地方（域）主义。

2.3 对赖特（Frank Lloyd Wright）的有机建筑和戴维·皮尔森（David Pearson）的新有机建筑的思考

2.3.1 两者理论比较

赖特的有机建筑理论	戴维·皮尔森的新有机建筑理论
1930年代提出	1980年代提出
1 强调建筑与自然的和谐统一 2 建筑自身的有机生长 3 相较于古典建筑的突破	1 强调为人而建造 2 有机建筑是自然能量流动的结果 3 不断融合又再生的轮回之旅
缺乏对人、自然及建筑关系的哲学探索	还有待进一步充实

赖特把有机建筑冻结和关入了静止的时空观里而没有进一步发展。这个在其城市作品的应对中显得与环境格格不入。本质都是一样：对自然的崇尚和对生命的礼赞。

2.3.2 希望它不是一个口号，不应停留在表层的形式和视觉效应追求上。

2.3.3 有机建筑理念整体上表现出一种"复归"的趋势，但这种复归不是简单的还原，而是在更高层次上的否定之否定的发展。

三、结语

工作室提出的"蔓设计"的概念是受外在及内在因素启发而产生的关于设计理念和工作方式的工作室层面的哲学总结和思考，在此很难说清楚这一总结是先后于行动。希望它的提出能够在现在如此多元和混乱的设计思潮中，为整个团队带来创作上更为清晰、自由并且坚定的指向，并使之成为设计理念的思想动力和源泉。它的内涵还有待同仁们进一步地去发展、丰富、充实和完善，并将它化作一习惯融入、蔓延到每个建筑师设计生活的全过程当中，最终做出具有较高完成度的、与固有环境一起构成有机物态的建筑作品来。

写出来是为了厘清世事，也是为了澄清自己的困惑，更是为了在纷杂的建筑江湖中不要乱了我们自己的方寸。把"蔓设计"梳理到这里，心中好像清爽了很多，感觉自己的职业生涯好像又要重新开始。

2017.8.30 于北京五路居

参考文献：
[1] 陶银骥，武斌等.简明现代西方哲学词典[M].成都：四川人民出版社，1988.
[2] 张若诗，庄惟敏.新有机建筑理念与建设性后现代主义哲学思想的关联研究[J]. 世界建筑，2016(11).
[3] [美]罗伯特·文丘里.周卜颐，译.建筑的复杂性和矛盾性[M]. 北京：中国水利水电出版社，2006.
[4] [英]布莱恩·爱德华兹.周玉鹏，宋晔皓，译.可持续性建筑[M]. 北京：中国建筑工业出版社，2003.

Organically-Permeated Design

Text / Wang Zhenjun

Keywords: tendril, permeate, organically-permeated design, organic, organic state and organic construct, natural field force, low-level and high-level construct, topological psychology, organic architecture and new organic architecture, sustainability of architecture

I. The Context in which the Concept is Proposed and Its Implication

1.1 The problem introduction – idea and method

For most architects, the professional life is just like this: they rush and rush so much all the day that they seem to be on the production line, drawings are released set after set, yet the city is sinking down in mediocrity day after day. However, besieged by all sorts of distractions, you can simply do nothing but stare blankly at the works collection of your idol, or wait for a flight.

Ever since the establishment of WZJ Studio in 2009, I have been imaging that I am able to get rid of the foresaid status from then on. I wish, instead of being indulged in undertaking conventional projects as usual, that I am able to enter a status of research and thinking preceding practice, in which I am able to be fully devoted to, and to take my time accomplishing my favorite projects, without being easily disrupted by stringent time and other factors, under such circumstances it is more likely for me to carry all my philosophical reflection on projects and architectural pursuit throughout the entire process of the projects, in a bid to accomplish works with relatively higher degree of completion one by one, leaving less regret behind. For this purpose, I have been considering whether there is a brief and refined term to follow. Meanwhile, I can clearly convey the appeal to my team, and even the clients.

This question is discussed by the staff of the Studio at a meeting on matters of principle. A colleague of mine suggests that a good design must be accomplished progressively in a relaxed and indulged status, yet it does not just mean a slow design. In the current design ecology in China, architects are not allowed to do design slowly in the first place. "Slow" becomes a topic of discussion, indeed, it is impossible for us to do our design work slowly, but we need to deliver our design that enables us to permeate or stretch the concept throughout the whole design process. After 20 years of speedy development in urban construction, buildings designed in snack-styled manner dominate every corner of the city. Meanwhile, the city requires us to come up with works of a higher level of completion. Likewise, our clients also hope that we deliver submittals that are more delicate than our previous works. In addition, we yearn, from the bottom of our hearts, to carve out works that are worthy of the times, in order that we could win respect in more fierce market competition in the future.

As for the "màn" in "màn 延" (that is, to permeate or stretch), there are three words, "曼", "漫" and "蔓", according to the authoritative Xinhua Dictionary, referring to the status of extension in the linear, planar and stereoscopic dimensions, respectively. It is unquestionable that the three-dimensional "man(蔓)" is the right character to pertinently express the idea of an architect.

The organic state and vitality presented in "蔓" (meaning a tendril vine as a noun, and to permeate in three-dimensional manner as a verb) are in perfect conformance with the concept of organic construct that the Studio has been following since establishment, that starts with the organic architecture theory represented by Frank Wright, and that has been constantly enriched.

By this token, "organically-permeated design" perfectly expresses the Studio's appeal in terms of concept on and attitude to architectural creation, indeed.

1.2 The implication of organically – permeated design ("蔓设计" mentioned in this book is the same with "蔓·设计" in the title)

1.2.1 Permeated – organic state and organic construct – a manner and attitude to understand view and intervene in nature, a design philosophy.

Indeed, this is by no means a new concept, but it is a theory that has been enriched constantly since Frank Wright. When we overlook every project site, or carefully observe the texture of a tree, a leaf and a stone, it is not difficult for us to find that they are full of harmony and unity of the identical natural force. All the techniques and art forms in the world can be derived from the existence forms of various creatures. However, humans always claim to have the perfect structure in the world, and demand everything in the world to be perfect, especially in the current time, when human capabilities have swollen uncontrollably, which makes the desire to pursue perfection even more vehement.

As an outcome of human activities, buildings fully reflect the laws governing this kind of human activity, and cities and buildings, as an artificial state of matter, have developed in human's pursuit for, and exploration of perfection. Unfortunately, buildings and cities, while displaying the strength and greatness of humans, also exhibit unsatisfactory status of development: as an artificial state of matter, buildings appear to be in contradiction and confrontation with natural environment, and the confrontation tends to exacerbate. Nature creates humans, who in return create buildings and modern cities and modern urban life. Modern life, unfortunately, alienates us from the nature more and more, which incurs countless misinterpretations of nature, and consequently, we miss the chance to grasp the essence of the energy.

French scientist Claude Allègre writes in *Ecologie des Villes, Ecologie des Champs* (*Ecology of Cities, Ecology of Countries*), "… The contemporary humans must understand that the era of mining has ended; what is before us is an era of management and preservation. The era of confrontation is over; what is ahead of us

Fig. 1: Natural Field Force

is an era of harmony."

(1) The organic construct concept in "organically-permeated design"

Undoubtedly, it is high time that humans need to earnestly examine the relations between architecture and nature. There is a surge of interest in "back to nature" in architectural philosophy, but we should be vigilant against two tendencies: one is that some architects mechanically apply the concept of "city with hills and waters" proposed by Qian Xuesen: as a result, hills and waters are applied indiscriminately regardless of the scale of the building; the other is that "green building" or "energy-efficient building" is labelled wantonly.

The organic construct idea in "organically-permeated design" advocates that we apply this pattern of getting along with nature in the all-around way, instead of imitating superficially (a square box might be organic at a certain place in a certain time) or filling it up mechanically, as nature itself is an extremely complicated and organic system and architects should intuitively employ human capacity to simplify and refine for the reasons of cost and emotion. Therefore, the "organically-permeated design" based on the organic construct concept should be a theoretical system that is future-oriented, open, and constantly enriched.

(2) The common features of the natural organic state and artificial organic state- diversified, organic, high-level, and systematic .

① The implication of organic: first it refers to a complex entity, the entirety of which is richer than every separate component, and each component is characterized by the feature that it would become organic just because it joins the entirety. So the typical attributes are entirety and property of life.

② Definition of organic state relations: each state of matter has its own distinctive individuality yet they co-exist in relation with certain factors, e.g. the organic state of matter presented by the natural state with the effect of various field force (including sunshine, wind, rain, snow, heat, freeze, gravity, and seismic force, etc.). It also refers to the law governing the self-development of an organic state of matter (that is birth, aging, death, and rebirth). (Fig. 1)

③ As a late-entrant, the artificial state of matter (buildings) should be related with the existing organic state of matter in the nature once it is involved in the nature; the quality of the relations depends on the architects: it requires the architects to leverage their knowledge, inspiration and wisdom to handle the relations, so that it maintains organic state relations, instead of stiff or isolated relations with the existing state of matter.

(3) Clarification of the concept of architectural sustainability

① This is not a slogan. As it primarily emphasizes thought and technology, it is an overall outlook on architecture, rather than the doctrine of an architectural school, as it does not stress image and style.

② Both the current moment and the future of the whole project life cycle should be taken into consideration.

③ It refers to the whole process of the project from planning throughout completion, not just the design stage.

④ You need always consider how to make full use of all the natural resources with the grace of nature.

1.2.2 Organically-permeated design — a way of working that persistently applies the design concept throughout the project

(1) Organically-permeated design is not slow design, but the design that applies the philosophy all the way round — the way and means of thinking and working that carries the design concept throughout the design process.

(2) Permeating the whole process: it applies throughout the whole process, in programming, planning, scheming for an individual building, preliminary design, shop drawing design, interior design, landscape design, field service, etc.

(3) Permeating all the means — scrutinizing and optimizing in overall modes, from planar to 3-dimensional, from plane drawings to 3-D models …

II. Analysis and Consideration

2.1 The relations between psychological factors and organic state of matter

2.1.1 The revelation of Topological Psychology.

When it comes to the question of the relations between the person and environment, we must refer to the equation $B=f(PE)$ developed by Kurt Lewin, a German psychologist who founded Topological Psychology. It states that behavior (B) is a function of the person (P) in their environment (E). Something attracts persons with attractive value (positive atomic value), which is what people are willing to approach and obtain; something rejects persons with rejective value (negative atomic value), which is what people are reluctant to accept or reject.

2.1.2 The issue that we should take into consideration in our design creation is how to meet men's psychological needs from the fundamental structural features of constructive relations of architecture as well as how to meet the needs of the complexity and uncertainty of state to matter relations reflected in psychological factors.The distinctive features should be stressed in local areas, but on the whole the focus should be on composition relationships and arrangement pattern, i.e. the

Fig. 2: National Center for Atmospheric Research (Architect: Leoh Ming Pei)
(Photograph: The National Center for Atmospheric Research, ©Copyright UCAR)

organic state relations in terms of the constructive relations of architecture.

2.2 Probing into the issues to be resolved today and tomorrow

2.2.1 The limitation of the square box in modernism and its way of disintegration

Humans have the instinct of simplifying and abstracting the natural state of matter, and this capacity has reached its apex thanks to technological advancement, but now we need to return to the pursuit for organic state of matter.

(1) The inevitability of the way of disintegration

The uniformity resulted from the "modern architectural principle" and the organic state of matter give us the revelation as follows: it is really necessary to introduce diversification into architecture, and the progress of technologies and thinking provide guarantee for the diversification. The unity brought about by various principles is contradictory with the diversification provided by the progress in technologies and thinking, and inconsistent with the abundant nature. Humans crave for being eternally engaged in rheological and nice organic state relationships.

(2) The questions arising from examination of organic state relationships

Organic state relationships result from the high-level and multi-directional combination of the natural field force, human capacity to transform the environment, and the field force generated by psychological factors. Based on this, architectural design should lay emphasis on individuality while being restricted by searching for the maintaining factors (various field forces) and reflecting on future development and change. Our creativity can be viable only when it is maintained in this field force. Mr. Leoh Ming Pei has set a good example for us when he designed the National Center for Atmospheric Research (NCRA) (Fig. 2). In order to understand the site, he used to stay the night with a sleeping-bag on the site, and drive around the site to investigate the Indian rock structure ruins nearby so as to draw inspiration for the design. Alvaro Siza's Boa Nova canteen (Fig. 3) based on the refining made upon the landform and micro-geographical formal composition that are special to the site, the plan, mass and rooftop shape stem from the careful study of the rocky cape site. His buildings give us a strong impression that they are the results of aggregation and interaction of the vectorial natural filed forces drawn from the place.

(3) Criticism on the trend of mechanical imitation of nature and traditional local architecture

Due to the backward technological level, traditional local architecture is a purely passive adaptation to the environment while yielding to natural field forces altogether, which needs to be improved through injecting the energy of contemporary science and technology. Surprisingly, some architects play tricks in design by using the raw methods of decades back, which is both a sheer waste of manpower and depletion of the land resources. Formalized imitation of natural state of matter is a misinterpretation of the organic state architecture all the more. Such inconceivable practice only displays the mechanical and low-level attitude on creation.

2.2.2 Thought on the development of Chinese architecture

(1) The constructive relations for current and future architecture should be aimed at developing active and positive organic state relations

The advanced technologies and highly-developed philosophy today enable us to dissect the generation, variation and development of various organic relations, and to actively thinking how to integrate them with architecture.

(2) Thought on shifting "field force elements"

Making use of the shift of the field force elements of different sites to create positive, active and abundant buildings with organic state relationships. It is unnecessary to worry about the convergence of design results, because the sites are varied and diverse although the same thinking principle is applied.

(3) Thought on high-level organic state relationships

Today people are capable of observing the intrinsic relations between things and their laws beyond the appearance, even reaching the detached state from the object to spirit. When applied on architectural creation, such capacity of humans means an in-depth reflection on the architectural constructive relations, i.e. it seeks for the organic state relationships influenced by the field force factors that generate high-level "effect of state".

2.2.3 Comments on Chinese architectural design

(1) The current chaotic phenomena in Chinese architecture. The primary service recipient of an architectural work is not the client any longer, but the public and the social media; every architect who tries to survive in the whirlpool of public opinions will be subject to tests; architecture has never received so much attention as today, yet nobody knows what this means. "… Architects are merely cash-generating machines of design styles… The architectural assessment system has become weird and elusive, architects can be wizards and Taoist priests combined, and the concepts upheld by architects are merely the alchemy for real estate…". "When numerous architectural theorists believe that the uncertainty of architectural form is the potential factor of new architecture, just as many fashionable factors will also be employed in architecture to interpret the significance of the times."

(2) Only Chinese can create the architecture of their own. We should strive to make

Fig. 3: Boa Nova canteen (Architect: Alvaro Siza)

Chinese architecture that has modern feel and that responds to the social changes of contemporary China from contemporary perspective. In terms of the creation in China's contemporary architecture, it is not rare that some architectural works "take the novelty of others as criterion", mimicking like a parrot. It is true that techniques and strategies are important; however, due to lack of concepts and principles, one will rest on the skills level eternally, and get eventually lost in the Western evaluation criteria. The responsibility of making contemporary Chinese architecture still falls on the shoulders of Chinese architects.

(3) The real innovation in architecture has not arrived yet, as no general knowledge has dominated us. However, the cornerstone of architecture remains steadfast, for we are in an era of language innovation and technique innovation only. Regionalism cannot resolve the universal problem, so it is a false regionalism.

2.3 Thought on Frank Lloyed Wright's organic architecture and David Pearson's new organic architecture

2.3.1 Comparison of the two theories

Organic architecture theory by Frank Lloyed Wright	New organic architecture theory by David Pearson
Founded in 1930s	Founded in 1980s
1. Stress on the harmony and unity between architecture and nature	1. Stress on building for people
2. Organic growth of buildings	2. Organic architecture is the result of natural energy flow
3. A breakthrough compared with classic architecture	3. A trip of recycling by continuous blending and rebirth
Lack of philosophical probing into the relations between human, nature and building	To be enriched further

Wright freezes organic architecture and locks it into static time-space view without any further development. This can be seen from his urban works, which seem incompatible with the environment. They share the identical essence: the worship for nature and the psalm for life.

2.3.2 Hope it is not a slogan, and one should not rest on the superficial form and pursuit for visual effect.

2.3.3 On the whole, the organic architecture concept takes on the tendency of "retainment", which is not a mere restoration, but the development of negation of negation on a higher level.

III. Conclusion

The concept of "organically-permeated design" proposed by the Studio is a philosophical summary of and reflection on design philosophy and way of working at the studio level inspired by internal and external factors, and it is difficult to make it clear whether the summary precedes or follows action. It is expected that this concept can give the whole team a more clear, free and firm direction in design amid the current diversified and bewildering design trends, making it the ideological drive and source for design concepts. Its connotation remains further development, enrichment, replenishment and improvement by colleagues. Architects are expected to develop a habit out of it and enable it to permeate the entire process of their design life. Ultimately they can deliver their architectural works that are of a high level of completion and that constitute organic state of matter together with the inherent environment.

Writing this out is for clarifying my train of thought, clearing my own confusion, what is more, for the purpose of not having our heart troubled in this confusing and disorderly architectural community. I feel greatly fresh and cool at heart as I have sorted out my ideas on "organically-permeated design". I feel that my professional career will start all over again.

Written in Wuluju ,Beijing on 30th, Augest, 2017

Bibliography:

1. Tao Yinbiao. Wu Bin. etc. *Concise Dictionary on Modern Western Philosophy* [M]. Chengdu: Sichuan People's Press, 1988.
2. Zhang Ruoshi, Zhuang Weimin. *The Connection Between the Concept of New Organic Architecture and the Philosophy of Constructive Postmodernism* [J]. World Architecture, 2016 (11).
3. Robert Venturi. tr. Zhou Buyi. *Complexity and Contradiction in Architecture* [M]. Beijing: China Water Power Press, 2006.
4. Brian Edwards. tr. Zhou Yupeng & Song Yehao. *Sustainable architecture* [M]. Brijing: China Architecture & Building Press, 2003.

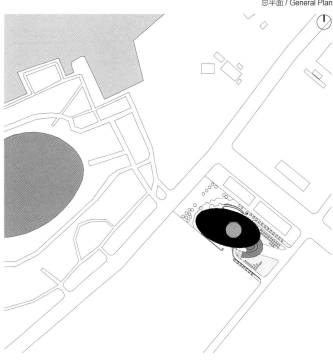
总平面 / General Plan

2010年上海世博会沙特国家馆
The Saudi Arabia Pavilion at Shanghai EXPO 2010
2007 ~ 2010 , Shanghai , 6100 ㎡

沙特馆是2010年上海世博会上唯一由中国人独立设计完成的外国馆。该设计用一种轻松的建筑语言描绘了阿拉伯神话中的月亮船沿着海上丝绸之路从阿拉伯半岛漂浮到东方时尚港口——上海的场景。极具动感和未来感的体量漂浮在地表之上，上扩下收的造型营造出大量凉爽舒适的室外等候空间和表演空间；室内外参观流线围绕中庭环形布置，外环上行内环下行的安排使进出人流舒缓有序，并且核心中庭充分利用自然采光和通风使建筑自成体系，达到节能环保的效果；空间展示上，创新性地采用全球独创的"全景融入式立体参观方式"，即以船体内壳作为展示投影屏幕，展廊架空其上，用极具体验性地融入式动线布局，使其展示效果和文化信息传递量最大化，似驾乘在阿拉伯魔毯上的参观体验，并与等候区阿拉伯风情的景观、开敞的歌舞表演平台以及视野极佳的屋顶花园，一起为游客奉献了一场汇集阿拉伯人文风情与地域文化的饕餮盛宴。沙特馆以清晰的创作概念、建筑形象以及与形象高度融合的参观方式获得了超高人气，很好地诠释有机建构理念的真正含义和价值所在。

The Saudi Arabia Pavilion at Shanghai EXPO 2010 is the sole foreign-built pavilion whose design is independently performed by a Chinese architect. The design depicts a scene of the moon boat in the Arabian mythology floating to the fashionable oriental port — Shanghai along the maritime Silk Road from the Arabia Peninsula by using a light-hearted architectural language. The extremely dynamic and futuristic mass stands adrift on the surface, and the V-shaped form creates a large area of cool and comfortable outdoor waiting space and performing space. The indoor and outdoor visitors' circulations are arranged around the atrium, giving ease and order to both the ascending visitor flow on the outer ring and the descending one on the inner ring; besides, the natural lighting and ventilation introduced in the atrium enables the pavilion to establish a system of its own, achieving the effect of energy saving and environmental protection. In terms of spatial exhibition, adopted is the world's first ever "panoramically syncretic 3D visiting mode", that is, the inner shell of the hull is used as the projection screen for exhibition, with the exhibition gallery mounted around the inner shell; the highly-experiential syncretic circulation arrangement maximizes the exhibition effect and convey of cultural message. The Saudi Arabia Pavilion, with its distinct concept for architectural creation, and architectural image, as well as the visiting mode that is highly compatible with the image wins great popularity among visitors, and serves as a good interpretation for the true implication and value of the organic construction concept.

首层平面 / First Floor Plan
三层平面 / Third Floor Plan

1 VIP 门厅
2 接待大厅
3 服务台
4 办公室
5 礼品处
6 服务间
7 储藏室
8 表演舞台
9 水景
10 设备间
11 观众入口
12 观众出口
13 VIP 出入口
14 地库出入口

1 设备间
2 多媒体技术间
3 备用间
4 展厅

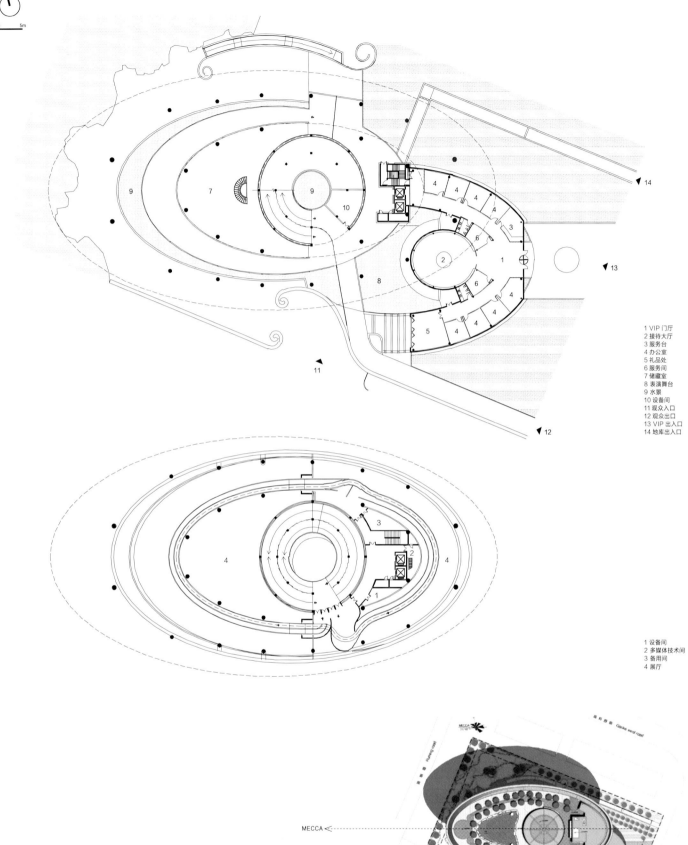

总平面 / General Plan

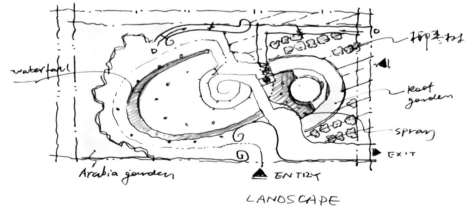

LANDSCAPE

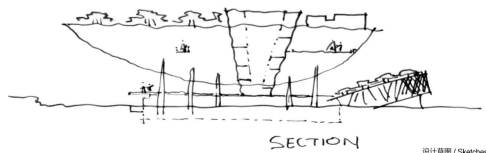

SECTION

设计草图 / Sketches

传统：看与被看二元并置　　　　　　创新：全景融入式观影模式

设计草图 / Sketches

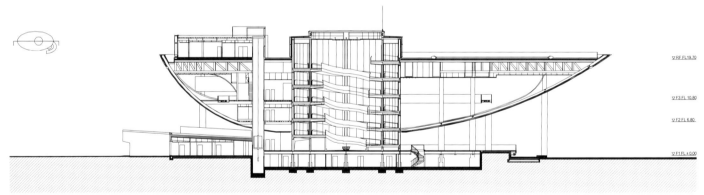

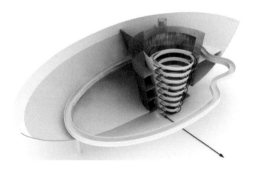

剖面 / Section
全景融入式展示空间 / Panoramic Integrated Exhibition Space

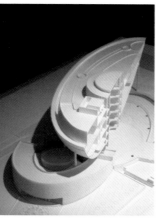
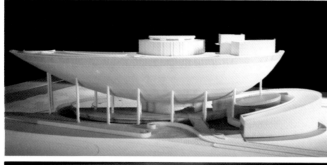

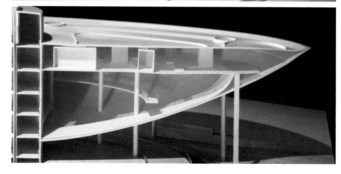

模型 / Models
建筑表皮与灯光效果推敲 / Deliberation on Building Skins and Lighting Effect

来自：穆罕默德·阿恩海姆迪教授
职衔：沙特馆执行官；沙特阿拉伯国王大学建筑系教授

From: Dr. Mohammad Alissan Alghamdi
Title: Executive director of Saudi Pavilion;
       Professor of Architecture and Building Science at King Saud University

来自：托马斯·赫斯维克
职衔：托马斯·赫斯维克工作室创始人、主持人；上海世博会英国馆建筑师

From: Thomas Heatherwick
Title: Founder and Director of Heatherwick Studio;
       Architect of UK Pavilion at Shanghai EXPO 2010

来自：克利福德·皮尔逊
职衔：《建筑记录》杂志副主编

From: Clifford Pearson
Title: Deputy Editor of Architectural Record

- 我们最终选择了王先生的方案，因为该设计运用现代建筑语言、技术和材料出色地诠释了阿拉伯文化和传统，因而在本次设计竞赛中脱颖而出，斩获竞赛头奖。

——穆罕默德教授

We ultimately accepted Mr.Wang's proposal because he was the first place winner of this competition as he excelled in his proposal in reflecting the Arab culture and traditions with modern archi-language, technology and materials.

—— Dr. Mohammad Alissan Alghamdi

- 从世博会访客和中外建筑界的评价来看，沙特馆无疑是成功的，这要归功于设计师的工作和付出。这件作品展示出的建筑造诣使其成为当今中国建筑师中的佼佼者。

——托马斯·赫斯维克

The success of the Saudi Arabia Pavilion both with visitors to the Expo and the architectural community in China and overseas is a credit to his work and commitment. In his work he has risen to the level of accomplishment that is shared only by a few Chinese architect working today.

——Thomas Heatherwick

- 我对于王振军先生将大胆的标志性结构与全方位多媒体室内展示高度融合的体验印象深刻。这个项目的成功使他跻身于世界级资质的精英建筑师队伍。沙特馆不仅成为了世博会最受游客欢迎的展馆之一，也在建筑师界及中外媒体的关注报道中广受好评。

——克利福德·皮尔逊

I was impressed by Mr.Wang Zhenjun's ability to integrate a bold iconic structure with a sophisticated multi-media interior experience. His success with this project places him in an elite group of architects with world-class credentials. Not only has the Saudi Arabia Pavilion became one of the most popular pavilions with visitors to the Expo, but it has garnered acclaim from architects and drawn the attention of publication in China and abroad.

——Clifford Pearson

同行评价 / Reviews

总平面 / General Plan

上海浦东软件园一、二、三期

Pudong Software Park, Phase I,II,III in Shanghai
1995 ~ 2003 (Phase I,II) 2005 - 2012(Phase III), Shanghai , 15.6 ha(Phase I,II) 44.7ha (Phase III)

上海浦东软件园位于上海浦东张江高科技园区的中心地带。软件园一、二期与三期之间有祖冲之路贯穿，二者遥相呼应并自成一体，将有机的"蔓设计"理念演绎出各自独具特色的形式，前者为井然秩序的学院派风格，后者则更现代简洁、自由开放，前后建成时间相距近十年。

软件园一、二期的长条形基地尺度（240m x 500m）和软件研发建筑尺度使环形车道＋中心庭院的规划结构显得顺理成章，而庭院中职工餐厅的绿化屋顶与来自主入口方向小于5% 的微地形景观有机地衔接在一起，形成了一个被放大、被立体化的并富有情趣的公共生态休闲空间，优质的景观可被环绕周围的研发建筑中的人们一览无余。主楼在入口处的二层架空处理，使城市空间、园区空间以及园区的核心中庭空间有机地联系在一起，既开放又富有层次。环形道路的交通系统有效地保证了园区中庭的安静和舒适。以上这些在有着24m限高、容积率大于1.5的园区空间中非常难得和珍贵。
软件园III期的规划在延续基地四边道路入口位置不变的前提下，打破了原有控规的中心十字路规划结构，采用环形主路将基地划分成不同规模的地块，就近将吕家浜的河水引入基地中心，形成环形主路加中心湖的规划结构。一方面有效地降低了驶入园区的车速，同时这种向心式的富有层次的布局能够为满足不同规模的企业入驻提供更加多样化的选择。中心湖区沿岸的公共性建筑以极为合理的半径为研发企业提供着服务，中心湖成为了休憩的中心、交流的中心、研发成功时庆典的中心，整个规划生动地诠释了"科技只是一种媒介，人与自然通过它而有机地联系在一起"的设计价值观。

Shanghai Pudong Software Park is located in the central area of Zhangjiang Hi-Tech Park in Pudong, Shanghai. Separated by Zhuchongzhi Road, Phase I and II of the Software Park echo with Phase III at a distance, yet each having its own features; the two parts interprets the "organically-permeated design" concept in distinct forms: the former presents itself in an orderly academic style, while the latter is more modern, concise, free and open, though the completion of the two components are almost 10 years apart.
Based on the rectangular site (240x500m) of Phase I and II of the Software Park and the dimensions of the software R&D buildings, the planning structure of a circular driveway plus a central yard is logical; the green roof of the staff cafeteria in the yard is organically linked with the micro-relief landscape with a slope of less than 5% from the main entrance, thus forming a public ecological leisure space that is magnified, three-dimensional and enjoyable. The superior landscape is within the field of view of the people who work in the ambient R&D buildings. The open space in the second floor over the entrance of the main building connects the urban space, park space with the core atrium space in the park, creating open and tiered space for the park. The road system of a traffic circle effectively ensures the quiet and comfortable atmosphere for the park atrium, which is very difficult and valuable for the park space that has a 24m height limit and floor area ratio of greater than 1.5.
While keeping the position of the access to the ambient roads intact, the planning for Phase III of the Software Park divides the site into blocks of various dimensions with the circular main road. Water from the Lvjiabang Creek nearby is introduced into the center of the site so that the planning structure is characterized by the circular main road and the central pond. On the one hand, this structure effectively reduces the speed of vehicles driving into the park; on the other hand, the centripetal and hierarchical layout provides diversified options for resident companies of various scales. The public buildings along the bank of the central pond provide services to the R&D businesses with a very reasonable radius. The central pond becomes a center of repose, a center of communication, and a center of celebration for the success of R&D projects. The planning gives a vivid interpretation of the design values that "science and technology is a mere medium, through which people and nature are organically connected."

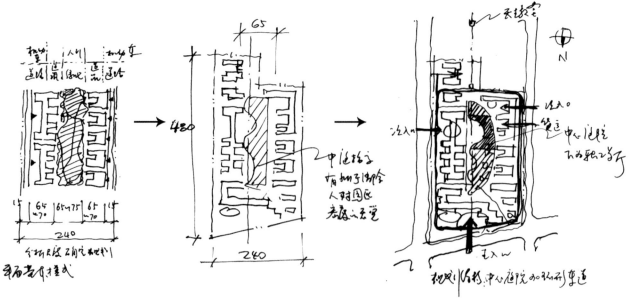

一、二期模型 / Model of Phase I and II
设计草图 / Sketches

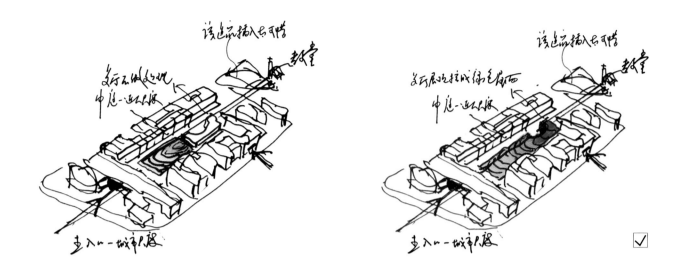

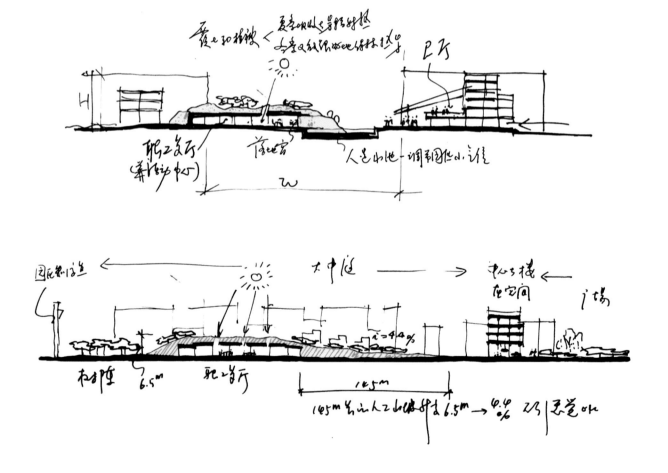

设计草图 / Sketches

三期模型 / Models of Phase III

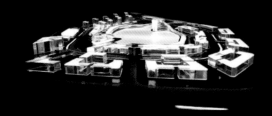

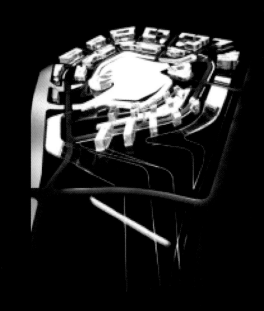

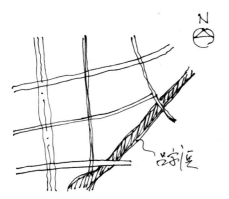
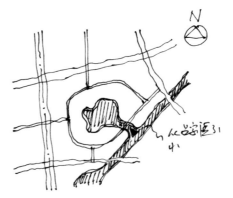

设计草图 / Sketches

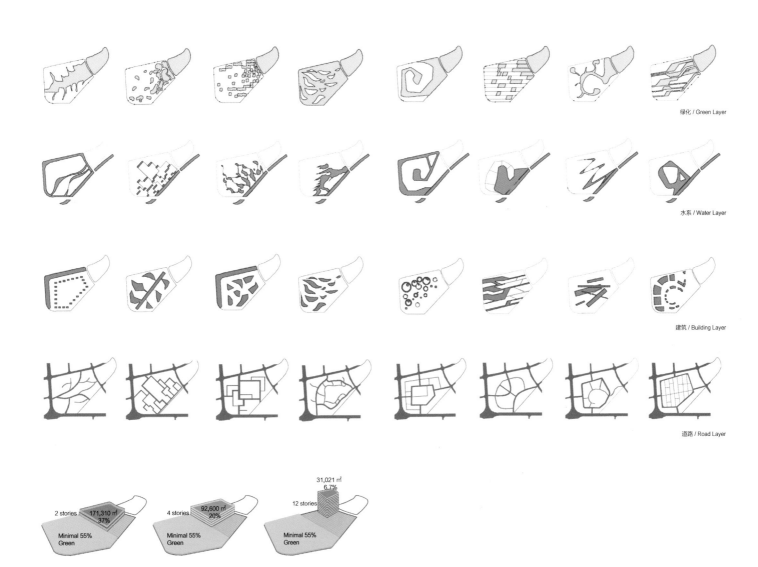

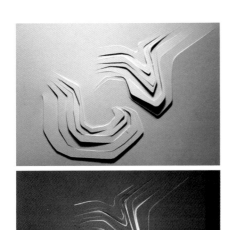

园区规划要素提取及研究 / Extraction and Study of Park Planning Elements
土地利用分析 / Land Ultilization Analysis
自然场力概念 / Concept of Natural Field Force

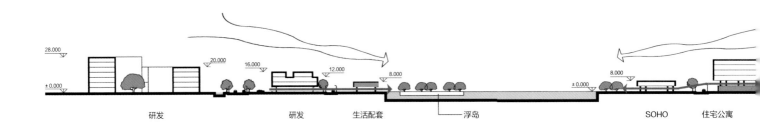

研发　　　　　研发　　生活配套　　　浮岛　　　　　　　SOHO　住宅公寓

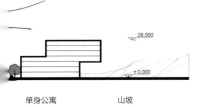

单身公寓　　　山坡

园区断面分析 / Section Analysis

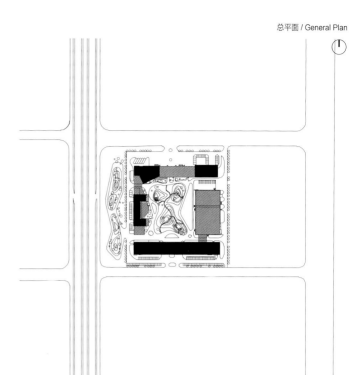

中国信达（合肥）灾备及后援基地

China Cinda (Hefei) Disaster Recovery and Back-up Base
2012 ~ 2016 , Hefei , 139800 ㎡

本项目为中国信达公司"二地三中心"灾备模式下的合肥中心建设项目，是以数据储存、查询处理、维护、传输、共享及综合分析等基本功能为核心的新型高科技综合体。基地由数据、后援、研发和后勤中心以"景观合院"形式组合而成。创作基于生态的有机建构理念，遵循"四水归堂"、藏风聚气等传统哲学思想，既有机地呼应了基地文脉也反映了地域特色文化，最大限度地整合利用外部环境资源的同时注重打造内部景观环境，使得整个建筑组群自身构建起良好的有机环境，将自然、人与建筑之间形成极度和谐、交融共生的生态秩序，以期为从事高强度的数据管理分析的研发和管理人员提供高质量的交流、聚会、相互激励的场所、激发灵感的场所。数据中心被置于最安全和安静的基地东侧，朝南侧以灰空间手法向城市打开形成主入口；高层研发楼置于基地北侧以减少对庭院的遮挡；后勤中心处在西侧城市小型街边公园与中庭的围合中，以柔性裙房与U形体块设计增加与中庭环境的接触面积。面向城市的一面继续采用连续界面手法将建筑组群有机地整合在一起，而将基地核心——数据中心处理成具有地域特点的新徽派建筑风格，以体现文化传承和人文关怀；基地制高点——研发大楼向主干道方向张开的同时与远处的巢湖风景区做出呼应。建筑外立面材料选用温润柔和的陶土板搭配简洁大方玻璃幕墙的方式，形成天然与人工、传统与现代、感性与理性的对比，二者刚柔并济。

It is the construction project of China Cinda Asset Management Corporation for its Hefei Center under the disaster recovery mode of "three centers built in two locations"; it is a new type of hi-tech complex focused on the basic functions of data storage, query processing, maintenance, transmission, sharing and comprehensive analysis. The Base consists of the building for data, backup, R&D and the logistics center in the form of a landscape courtyard. Based on the concept of ecologically organic construction, and complying with the traditional philosophical principles of "the rainwater of a courtyard flowing from the roofs into the yard", as well as hiding wind and gathering gas, the design organically responds to the context of the Base and reflects the culture with local features, focusing on constructing the internal landscape environment while maximizing the integration and utilization of the external environmental resources, so as to create an ecological order that is characterized by the extreme harmony among nature, man and building. It aims to provide a workplace for the R&D and management personnel engaged in data management and analysis, a sector involving intensive effort and high accountability, to enjoy high-quality communications, to have mutual stimulation and to activate the inspiration for R&D.The data center is placed on the east side of the site, which is the most secure and least disturbed zone, and it is opened to the city towards the south side with gray space to form the main entrance; the high-rise R&D building is placed on the north side of the Base in order to reduce shade over the courtyard; the logistics center is located on the west side, where the small local park and the courtyard enclose, the design of flexible podium and U-shape mass is applied to increase the contact area with the courtyard environment. The approach of continuous interface is introduced to organically integrate the cluster of buildings for the side facing the city. The building for data center ——the core of the base, takes advantage of the style of New Huizhou-style Architecture with the local geographical characteristics, so as to express the regional cultural inheritance and humanistic care in architecture; the R&D building ——the commanding height of the base, responds to the scenic zone of Chaohu Lake in the distance while opening towards the artery. The combination of mild & gentle ceramic plates and simple & elegant glass curtain walls are selected as the facade materials for the buildings, so that a contrast is visible between the natural and the artificial, the traditional and the modern, and the sense and sensibility, coupling hardness with softness.

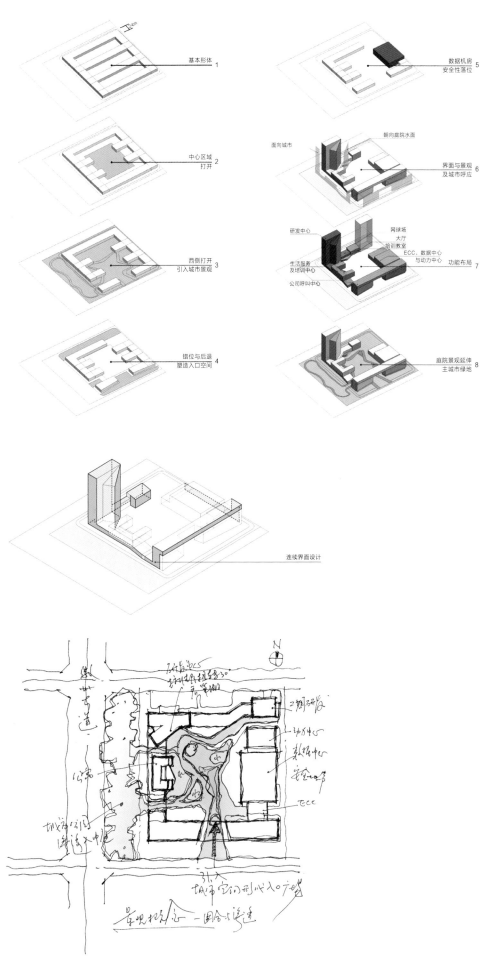

概念生成 / Concept Generation
建筑形体及材料建构概念 / Concept of Building Form and Material Construction
景观概念草图 / Sketch of Landscape Concept

立面 / Elevation
剖面 / Section
0 5m

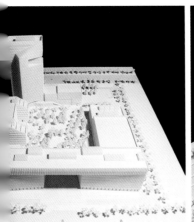
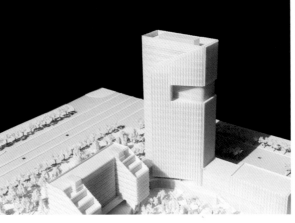

模型 / Models

首层平面 / First Floor Plan
三层平面 / Third Floor Plan
标准层平面 / Typical Floor Plan

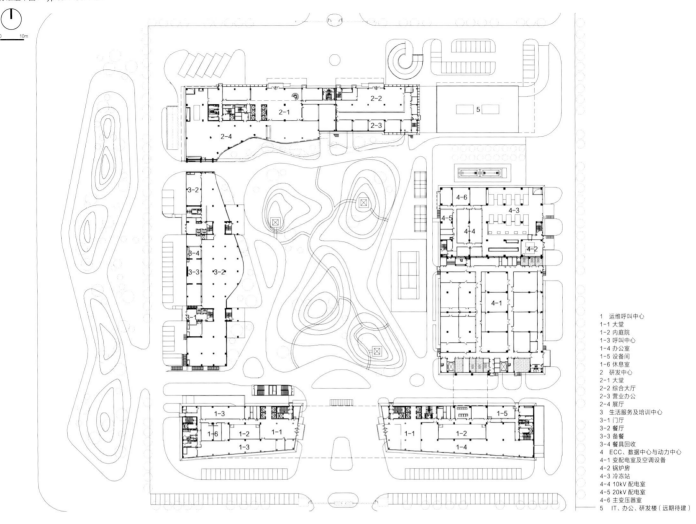

1　运维呼叫中心
1-1　大堂
1-2　内庭院
1-3　呼叫中心
1-4　办公室
1-5　设备间
1-6　休息室
2　研发中心
2-1　大堂
2-2　综合大厅
2-3　营业办公
2-4　展厅
3　生活服务及培训中心
3-1　门厅
3-2　餐厅
3-3　备餐
3-4　餐具回收
4　ECC、数据中心与动力中心
4-1　变配电室及空调设备
4-2　锅炉房
4-3　冷冻站
4-4　10kV配电室
4-5　20kV配电室
4-6　主变压器室
5　IT、办公、研发楼（远期待建）

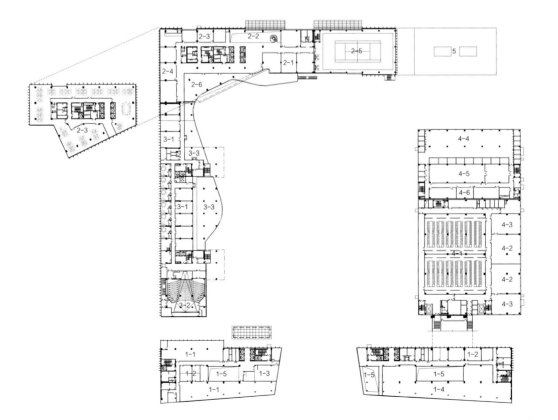

1　运维呼叫中心
1-1　呼叫中心
1-2　会议室
1-3　更衣室
1-4　办公区
1-5　上空
2　研发中心
2-1　培训教室
2-2　上空
2-3　办公室
2-4　设备间
2-5　室内网球场
2-6　屋顶花园
3　生活服务及培训中心
3-1　培训教室
3-2　阶梯教室
3-3　屋顶花园
4　ECC、数据中心与动力中心
4-1　机房模块室
4-2　UPS配电室
4-3　电池室
4-4　柴油发电机房
4-5　变配电室
4-6　应急备用间
5　IT、办公、研发楼（远期待建）

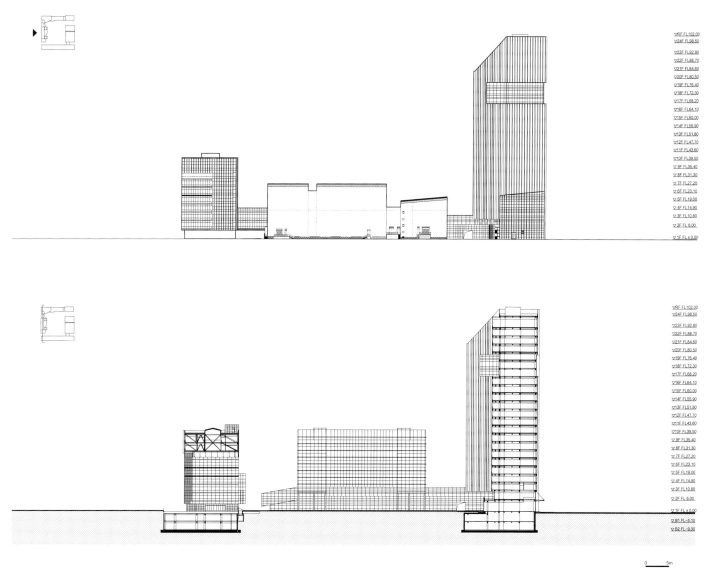

立面 / Elevation
剖面 / Section

建筑细部 / Detial View
研发中心细部 / Detial Drawing

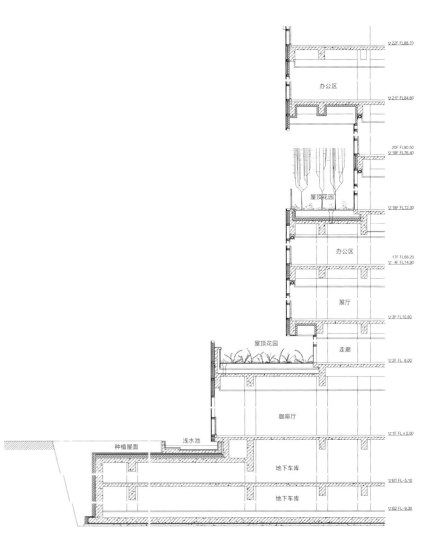

总平面 / General Plan

长沙中电软件园总部大楼

The Headquarters Building of Changsha CEC Software Park
2009 ~ 2013 , Changsha , 57600 ㎡

长沙中电软件园位于长沙麓谷国家高新技术产业开发区园内。其中总部大楼位于城市干道交叉口的西南角、西侧面向规划园区的中心景观湖面，自然环境优越宜人。总部大楼作为软件园内首幢标志性建筑，对整个园区的启动和发展起着至关重要的作用。设计因地制宜，在面向不同城市要素的界面做不同的表情处理，在两个方向、空间和形态上营造建筑与周边环境的和谐关系。将高耸的主楼置于面向城市主干道一侧，挺拔端庄而富有仪式感，面向园区景观湖的一侧用半圆环形裙房拥抱中心湖景。为使80米限高下的主楼呈现出更具向上的张力，在其顶部和底部采用了倾斜10°的处理手法；为适应南方炎热的气候特点，立面选择内凹式开窗组合的竖向肌理，配以土黄色陶土面砖，进一步加强了建筑纵向的力量感。临湖一侧弧线形的职工餐厅造型上采用外廊格栅式处理，突出建筑与中心湖景的渗透和融合，在遮阳的同时使建筑更具层次感。建筑整体注重传达湖湘文化大方、朴实和求变创新的精神气质。

Changsha CEC Software Park is located in Changsha Lugu National Hi-Tech Industry Development Zone. The headquarters building is on the southwest corner of the intersection of arterial streets, with the west side of the building facing the central waterscape lake in the planned park; the natural environment is attractive and delightful. As the first icon in the Software Park, the headquarters building plays a vital role for the startup and development of the whole Park. In the light of the specific conditions the design adopts different facial expressions for the interfaces facing different urban elements in a bid to create harmonious relations between buildings and their ambient environment in terms of space and shape in two directions. The soaring main building on the side facing the urban artery looks lofty and dignified with a sense of rite; the side facing the waterscape lake in the park embraces the central lakeview with its semi-circular podium. In order to make the main building with 80-meter limit appear to be more towering a 10-degree inclination is applied in both the bottom and top of the building. The vertical texture created by the indented windowing in consideration of the sweltering climate in south China, combined with the earthly yellow clay tiles, reinforces the sense of vertical strength of the building. The grating veranda outside the arc-shaped staff cafeteria on the side facing the lake highlights the infiltration and fusion between the building and the central lakeview, giving the building an extra layer while providing sunshade. The entirety of the headquarters building lays emphasis on convey the ethos of generosity, simplicity and innovation.

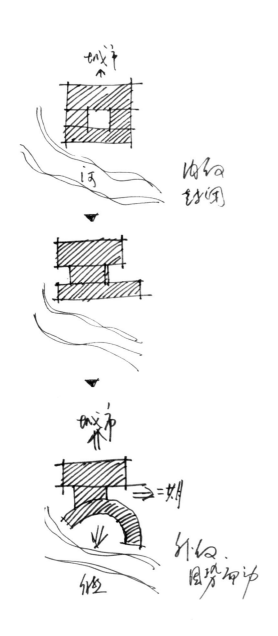

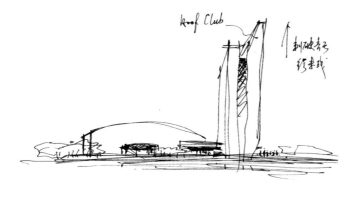

设计草图 / Sketches

首层平面 / First Floor Plan
二层平面 / Second Floor Plan
标准层平面 / Typical Floor Plan

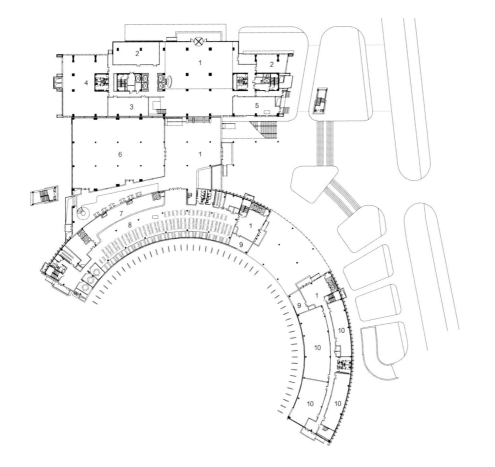

1 门厅
2 商务中心
3 档案室
4 咖啡厅
5 消防控制室
6 展厅
7 备餐区
8 职工餐饮
9 商业服务
10 技术中心

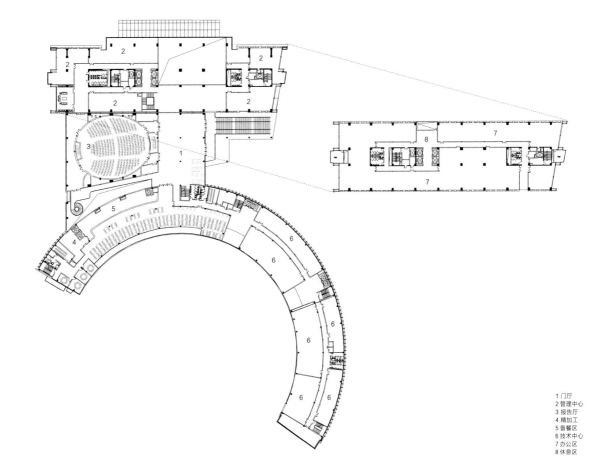

1 门厅
2 管理中心
3 报告厅
4 精加工
5 备餐区
6 技术中心
7 办公区
8 休息区

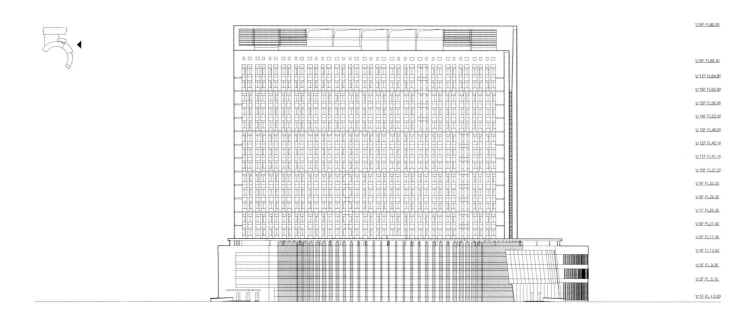

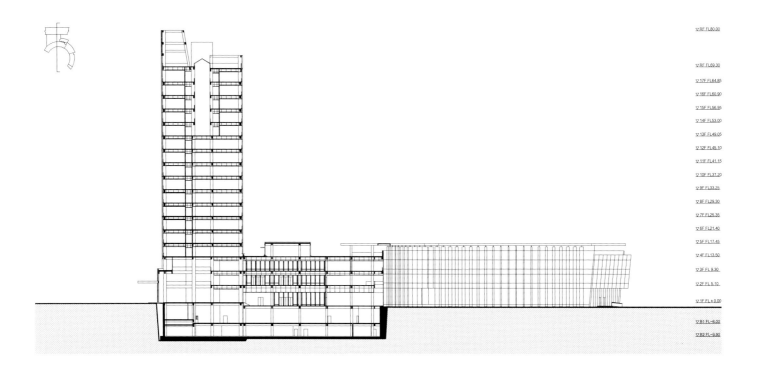

立面 / Elevation
剖面 / Section

模型 /Models

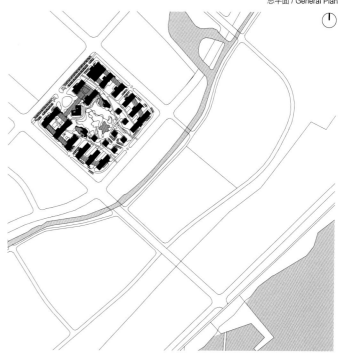

总平面 / General Plan

海盐杭州湾智能制造创新中心
Haiyan Intelligent Manufacturing Innovation Center
2014 ~ 2018 , Zhejiang , 27.50 ha

这可能是中国距大海最近的园区，基地位于国家中欧合作试点的海盐县，临近滨海公园，距离杭州湾不到 800 米。规划以"交融"为概念，希望城市、海洋、人、科技与自然在此交织，最终交融在自然的脉络中。建筑高度和体量由北向南朝大海方向逐渐降低和缩小；环形主干道加组团支路的交通系统为中心景观湖实现纯步行化提供了保障；单体建筑将交通及辅助空间作模块化处理，与主研发楼体量组合在一起，以期为园区建设阶段的产业化提供可能。作为园区中心和制高点的总部大楼，以 V 字形平面面向大海方向，试图争取最大化的观海面，并使北侧自然形成欢迎之态的主入口广场空间；而空间和形式的处理则通过左右两个体量的错动、穿插、咬合等手法追求建筑的动态平衡和左右两个体量关系的相互协调，同时寄予"合作、信念、信心"等美好的期许。总部大楼成为园区的标志性建筑，更是观享海景的最佳位置。

This might be the science park that is closest to the sea in China as the site is less than 800 meters from Hangzhou Bay. The science park is situated in Haiyan, where the National Sino-Europe Cooperation Pilot Program is carried out, and the coastal park is nearby. With the concept of blending as the guiding principle, the planning aims to interweave the city with sea, people, science and nature, with all elements finally interlacing into the natural context. The building height and mass gradually descends from the north to the south towards the direction of sea. The traffic system consisting of the circular main road plus access roads to building clusters ensures that only sidewalks are available along the central waterscape pond. Modularization of traffic and supporting space is done for individual buildings. They are combined with the main building for R&D. With all these buildings in place, it is intended to ensure the industrialization for the park even during the construction phase. As the center and commanding height of the park, the headquarters building faces the sea with the V-shaped sector in an attempt to have a maximum view of the sea; meanwhile, a greeting square space is formed naturally at the main entrance on the north side. For the design of space and shape, the application of such approaches as dislocation, interspersing and snap-in of the two masses on the left and right side seeks a dynamic balance of the buildings and inter-coordination between them. The relationship of the two interactive buildings also implies the nice expectation of "cooperation, confidence and faith". The headquarters building is the icon of the park and also the optimal place to observe the seascape.

概念生成 / Concept Generation

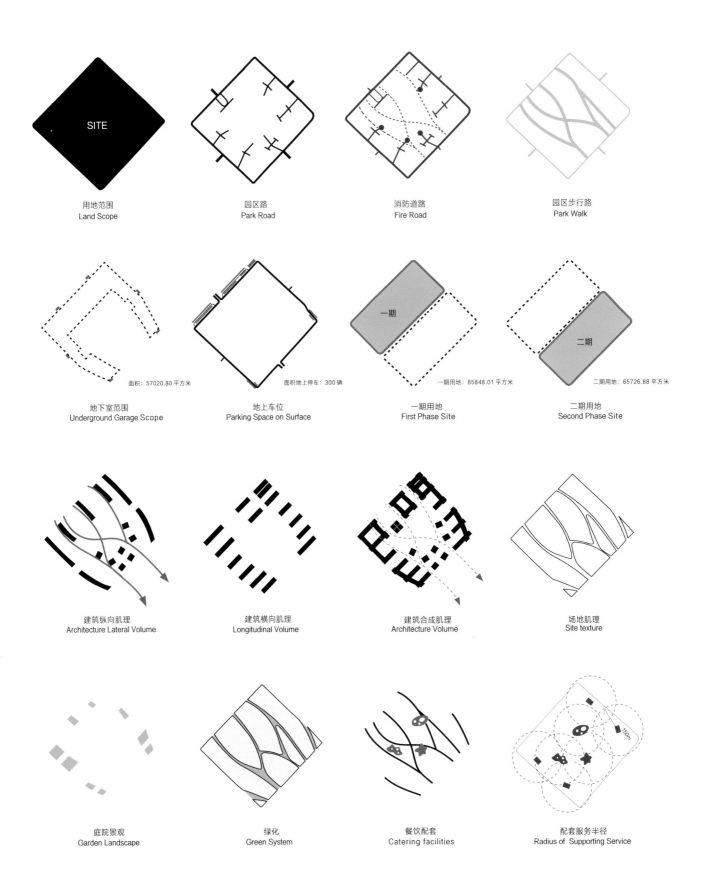

80

研发办公楼结构封顶

研发办公楼外立面装饰施工中

配套餐厅施工中

地下车库引入自然光

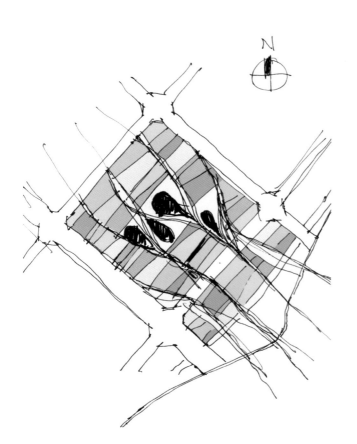

设计草图 /Sketch
施工实景/ Views of Construction

首层平面 / First Floor Plan
二层平面 / Second Floor Plan
标准层平面 / Typical Floor Plan

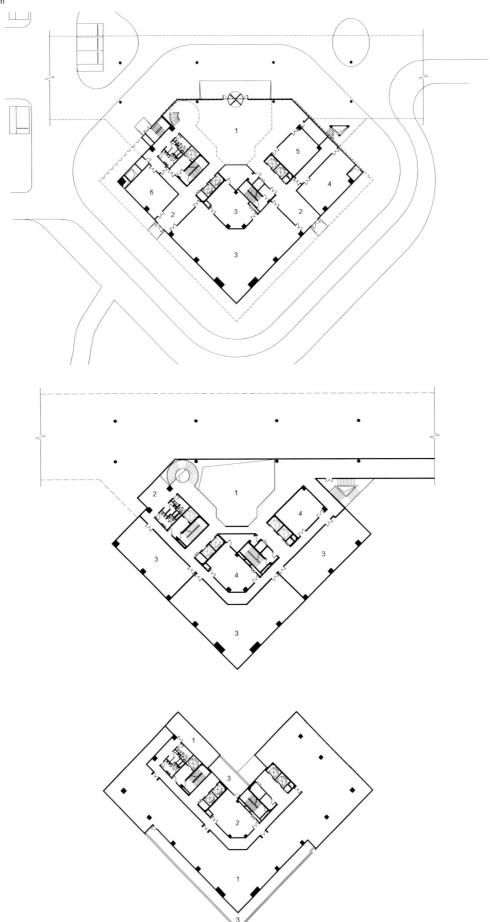

1 大堂
2 门厅
3 展厅
4 接待
5 商务
6 休息厅

1 大堂上空
2 休息区
3 研发办公区
4 会议室

1 研发办公区
2 会议室
3 屋面

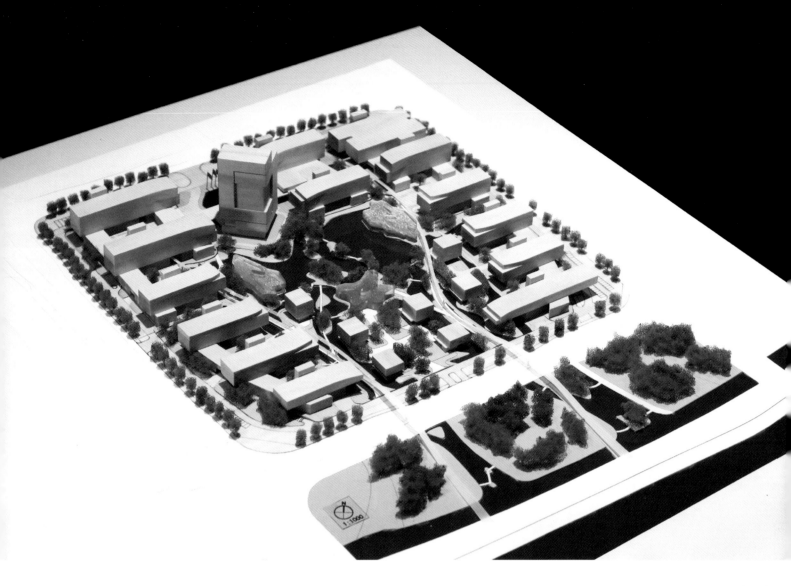

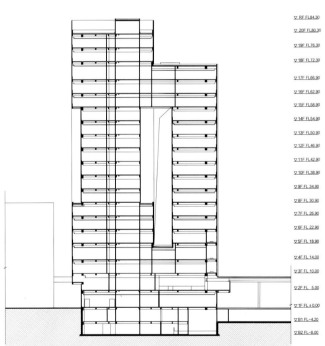

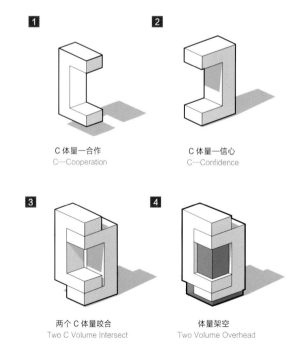

1 C 体量—合作 / C—Cooperation
2 C 体量—信心 / C—Confidence
3 两个 C 体量咬合 / Two C Volume Intersect
4 体量架空 / Two Volume Overhead

模型 / Model
剖面 / Section
总部大楼体量生成 / Volume Generation

幕墙材料：陶板（红褐色）

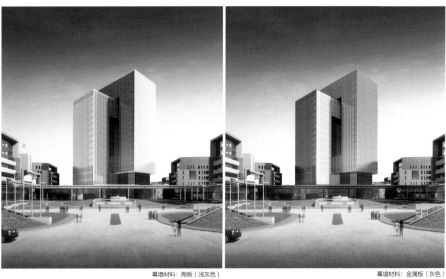

幕墙材料：陶板（浅灰色） 　　幕墙材料：金属板（灰色）

幕墙材料选择 / Curtain wall material Selection

首层平面 / First Floor Plan
二层平面 / Second Floor Plan
三层平面 / Third Floor Plan

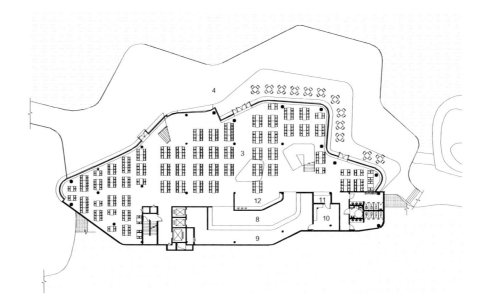

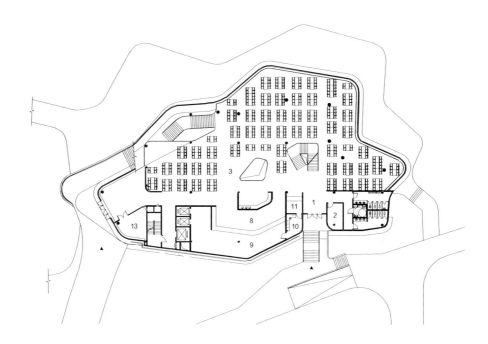

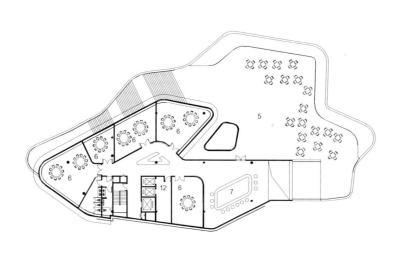

1 门厅
2 餐卡充值
3 就餐区
4 亲水平台
5 屋顶平台
6 餐厅包间
7 酒水区
8 取餐通道
9 送餐通道
10 洗碗间
11 收残区
12 备餐间
13 超市

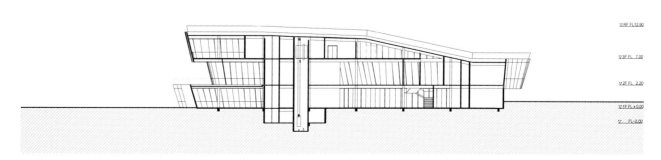

模型 / Model
剖面 / Section

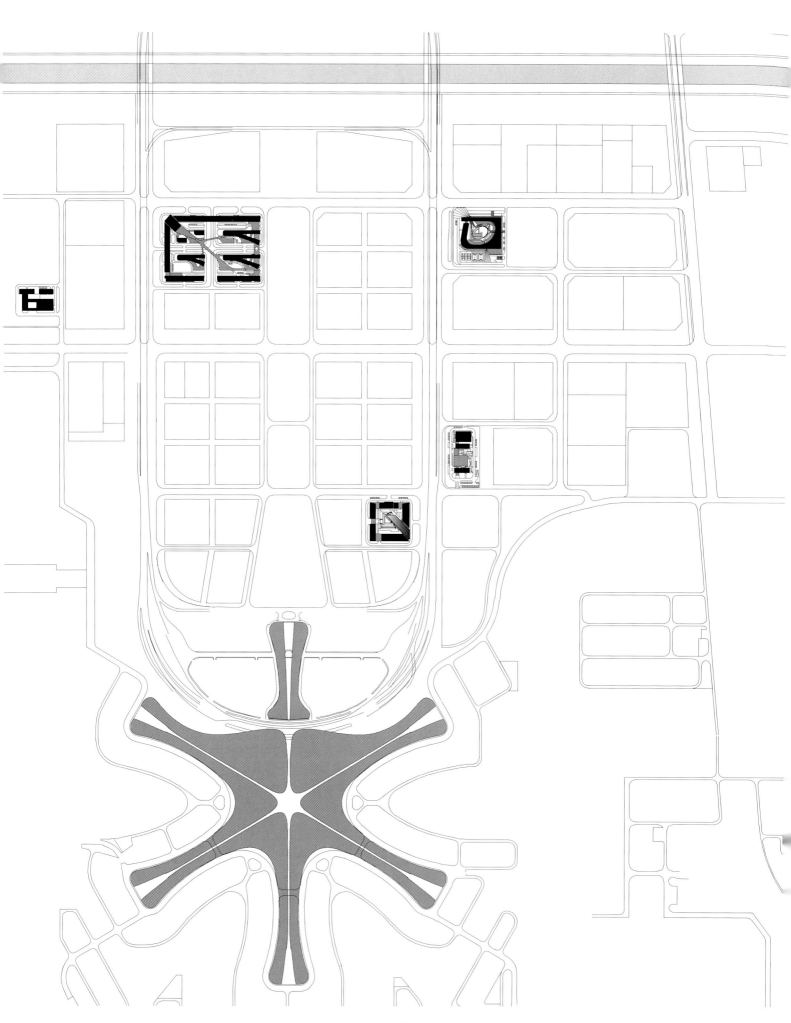

总平面 / General Plan

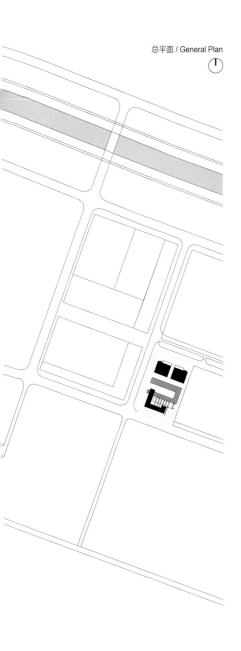

北京新机场系列项目
Beijing New Airport Series Projects

位于北京市大兴区的北京新机场定位于"综合性超大型国际机场",其相关附属设施的建设正是为全力打造新机场提供全面的配套设施与服务支持。现工作室已承接六项配套工程,分别是北京新机场信息中心(ITC)和指挥中心(AOC),行政综合业务用房工程、生活服务设施工程、非主基地办公项目、工作区物业工程及工作区车辆维修中心工程。设计方案均秉承"蔓设计"理念,意在营造具有生态、创新、便捷及前瞻性的复合型空间环境;在充分研究发展新机场的场所精神的同时,深入探索空港建筑的类型学特色,通过规划、建筑及景观语言的独创性表达,全面彰显新机场的建筑气质。

Located in Daxing District, Beijing New Airport is positioned as a "comprehensive super-large international airport", and the construction of relevant auxiliary facilities is intended to provide overall supporting facilities and services for making the new airport. Our studio has undertaken the design for six supporting projects as follows: The ITC and AOC for Beijing New Airport, Administrative Integrated Business Building Project, Living Service Facilities Project, Non-main-base Office Building Project, Working Area Property Management Building Project, and Working Area Vehicle Maintenance Center Project. The design schemes all adhere to the concept of "organically-permeated design", aiming to build composite spatial environment that features ecological, innovative, convenient and prospective traits. While we take great efforts in researching the spirit of place in the development of the new airport, in-depth exploration is also made into the typological features of airport architecture; it is aimed to fully display the architectural ethos of the new airport through a unique expression of the planning, architectural and landscape language.

北京新机场信息中心和指挥中心
The ITC and AOC for Beijing New Airport
2015 ~ 2019 , Beijing , 29998 ㎡

本项目以北京新机场所处的显赫位置和本身具备的重要枢纽功能，成为新机场区域内仅次于航站楼的最具标志性建筑。设计试图用"信息链"的概念将信息中心（ITC）与指挥中心（AOC）链接起来，将技术、自然与人结合起来，向大家展示出该建筑智能、有机、人文的本质特性和大气、理性、质朴、可靠的建筑性格。建筑西侧尽可能退让紧邻的机场高架，将以指挥办公为主的AOC部分置于地块南侧，以获取最好的景观和朝向；以数据运维为主的ITC部分置于临近支路的北侧，二者相对独立并通过架空通廊和景观步道串联成一个整体；办公和研发体量以咖色陶板材料塑造出连续链接环绕的体量节奏，强调安全的核心功能部分用返璞归真的清水混凝土板包裹嵌入，二者搭配散发出独特的人文气息；不同类型景观庭院的穿插其间，强调了高智能建筑崇尚自然以及对"以人为本"的设计原则的不懈诉求。ITC机房部分按照国际标准适度超前设计，一次施工、分部安装，便于后期扩展和机房等级提升，以充分考虑建筑全周期节能的可持续性。

Considering the outstanding position of this project in Beijing New Airport and its important pivotal function for the airport, it becomes a most iconic building, second only to the terminal, in the new airport area. The design attempts to link the ITC and AOC via the concept of information link, connecting technology, nature and people, so as to demonstrate the essential features of the building as being intelligent, organic, and people-oriented, and the architectural traits of being grand, rational, pristine and reliable. The west side of the building sets back as much as possible from the airport flyover nearby; the AOC, which is focused on command and office, is arranged on the south side of the plot so as to get the best view and orientation; the ITC, which is focused on data operation and maintenance, is placed on the north side, which is close to the access road, the two relatively detached parts are linked as a whole through elevated breezeway and landscape footpath; the coffee-colored ceramic plates applied on the office and R&D part generates the mass rhythm surrounded by continuous links; the core function part, which stresses security and safety, is embedded after being packaged with plain fair-faced concrete slabs, the collocation of the two materials gives off special cultural air; various types of landscape courtyards interleave among them, highlighting the advocating of nature in intelligent building as well as the unremitting appeal for the people-oriented design principle. Moderately forward-looking design is applied for the ITC mechanical room based on international standard, the construction of which is undertaken at one time while installation will be completed in steps, so as to facilitate subsequent expansion and upgrading of the mechanical room; thus, full consideration is given to the sustainability of energy efficiency in the full period of construction.

首层平面 / First Floor Plan
二层平面 / Second Floor Plan

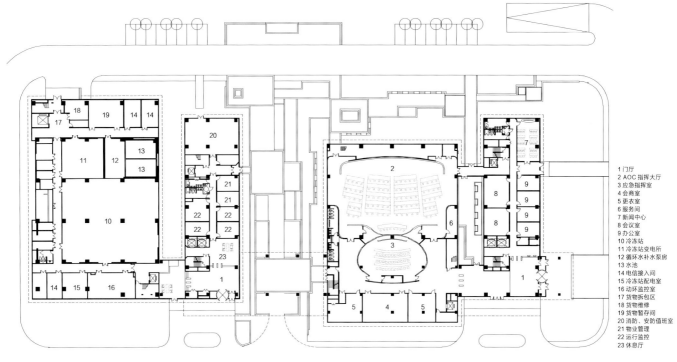

1 门厅
2 AOC 指挥大厅
3 应急指挥室
4 会商室
5 更衣室
6 服务间
7 新闻中心
8 会议室
9 办公室
10 冷冻站
11 冷冻站变电所
12 循环水补水泵房
13 水池
14 电信接入间
15 冷冻站配电室
16 动环监控室
17 货物拆包区
18 货物维修
19 货物暂存间
20 消防、安防值班室
21 物业管理
22 运行监控
23 休息厅

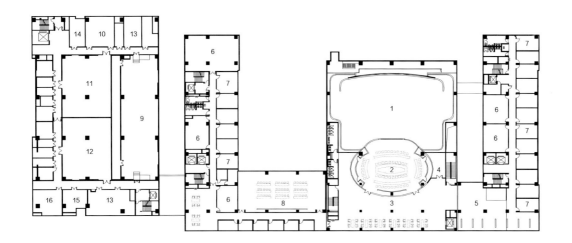

1 上空
2 综合会商室
3 休息区
4 服务间
5 展厅
6 会议室
7 办公室
8 系统运行监控室
9 测试机房
10 汇聚机房
11 10kV 变配电所 A
12 10kV 变配电所 B
13 UPS 电池间
14 备用间
15 脱机介质库
16 磁带库

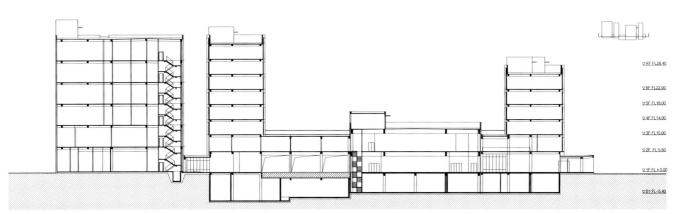

剖面 / Section

概念生成 / Concept Generation
模型 / Models

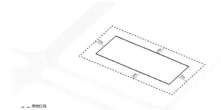

Step 1 项目用地：建筑项目规划用地 19950 ㎡，由于红线内有管线，建筑退让用地红线 20m。

Project land use: 19950 ㎡, the construction sets back 20m from the property line as there are pipelines underneath.

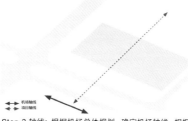

Step 2 轴线：根据机场总体规划，确定机场轴线。根据机场轴线，确定项目轴线。

Axis: The airport axis is determined by the master plan of the airport. The project axis is determined as per the airport axis.

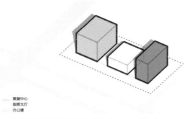

Step 3 功能布局：根据任务书要求，布置数据中心、指挥大厅和办公楼体块。

Functional layout: The ITC, AOC and office building masses are arranged as per the requirement of the project program.

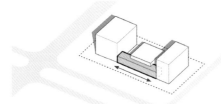

Step 4 连接：通过休息及共享空间连接数据中心、指挥大厅与办公空间。

Connection: The ITC, AOC and office space are connected through rest and shared spaces.

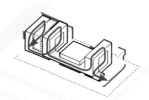

Step 5 信息链：概念植入建筑形体中。

Jetting-out: The architectural shape jets out towards the airport axis and airport, introducing the airport landscape while enriching the architectural form.

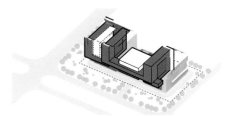

Step 6 完成：置入道路与绿化，完成项目。

Completion: embedding roads and greening to conclude the project.

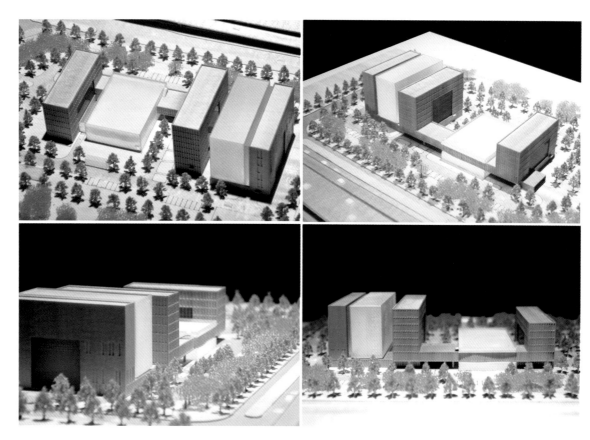

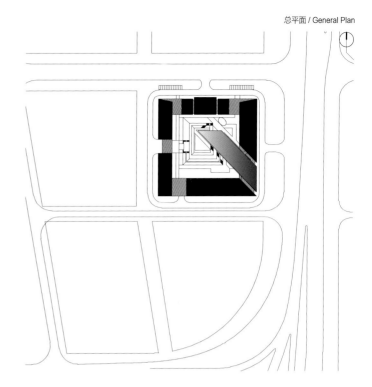

北京新机场行政综合业务用房工程

Beijing New Airport Administrative Integrated Business Building Project
2016 ~ 2019, Beijing , 69000 ㎡

本项目位于北京新机场出场高架的特殊位置上，是旅客往来机场必经的一栋建筑。设计以"北京门户"的概念为依据来呼应其所处的显要位置和重要功能。行政综合业务用房的两个办公单位呈"U"形组合，相对独立又在建筑形象上相互连接，与地块西侧待建的业务办公衔接围合成有机的整体；同时，提取出建筑群中展示、服务等公共性功能，采用体块介入、虚实对比的造型手法，在规则合院造型中斜向嵌入具有视觉冲击力的异形玻璃体量，形成建筑焦点，以轻巧、通透而又坚实有力的形象营造出一种崛起之势；围合的院落使建筑最大限度地拥抱景观环境，斜插的通透玻璃体也起到进一步放大和激活景观空间的效果；建筑主体表皮选择素雅大气的天然石材，与体现现代科技感的菱形玻璃幕墙材料相得益彰，展现出新机场与时俱进的前瞻性设计；综合管理、业务办公楼与工程档案、展示馆和服务中心多种功能呼应融合，互补共生，成为新机场的重要展示窗口。

As the project is located in a special position, by the exit flyover from Beijing New Airport, it is a building that passengers inevitably pass by while arriving at and departing from the airport. Based on the concept of "Beijing Gateway", the design highlights its prominent position and crucial functions. The two office blocks of the administrative integrated business building, in a U-shaped combination, are both relatively independent to each other and mutually connected to make an organic whole together with the business office building to be constructed on the west side of the plot. Meanwhile, the public functions such as exhibition and service are extracted from the building group; modeling tactics like block intervention and virtual-real contrast are applied when a visually-shocking irregular glass mass is obliquely embedded into the regular courtyard to form the architectural focus, creating a posture of rise with the light, transparent yet solid and forceful image; the enclosed courtyard enables the buildings to embrace the landscape environment to the maximum extent, while the obliquely cut-in transparent glass block help achieve the effect of further magnifying and activating the landscape space; simple but elegant natural stone is selected as the skin of the building, benefitting a company with the shape portraying the sense of modern technology and the diamond glass-wall material; all together, they indicate the forward-looking design of the new airport in a bid to keep pace with the times. The integrated management and business office building echoes the multiple functions of the project archive and exhibition pavilion, and service center, complementing each other to become an important display window for the new airport.

首层平面 / First Floor Plan
二层平面 / Second Floor Plan

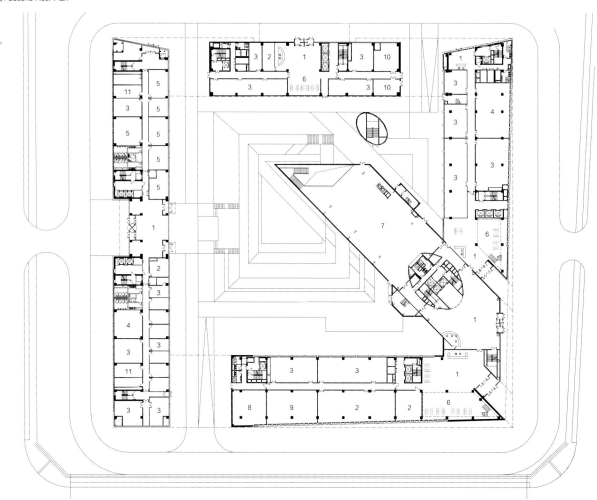

1 门厅
2 接待 / 门卫
3 办公室
4 会议室
5 培训教室
6 休息区
7 餐厅
8 商业
9 银行
10 安防 / 消防控制室
11 设备机房

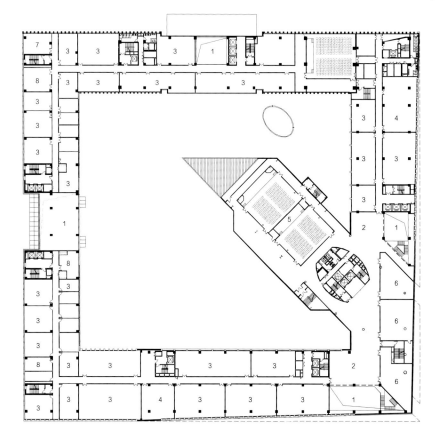

1 上空
2 公共区
3 办公室
4 会议室
5 报告厅
6 档案室
7 阅览室
8 设备机房

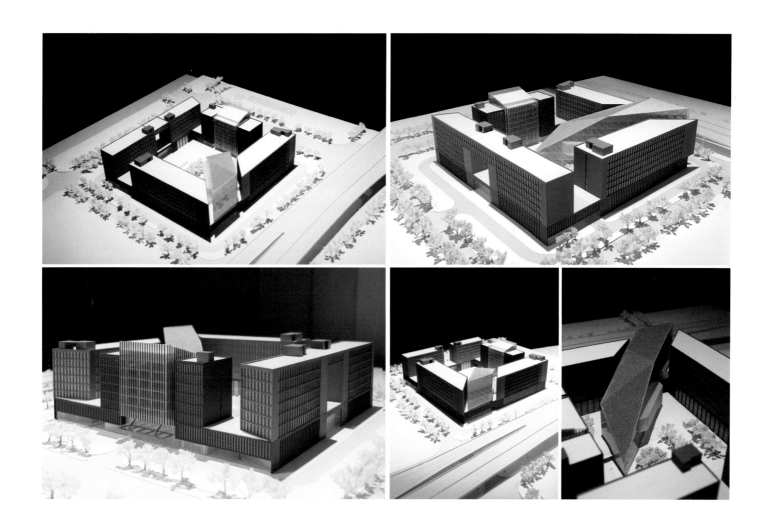
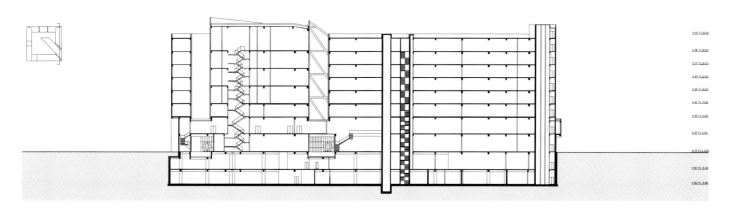

模型 / Models
剖面 / Section

总平面 / General Plan

北京新机场生活服务设施工程
Beijing New Airport Living Service Facilities Project
2016 ~ 2019 , Beijing , 66000 ㎡

本项目提出"绿色·交融"的设计概念，希望为使用者提供一个放松、便捷的复合型空间环境。设计中尝试将倒班宿舍、生活服务、职工食堂等功能重新组织，以开放与围合的方式进行垂直分区。其中下部生活服务中心及职工食堂部分引入开放街区的概念，为机场6000人提供便捷的就餐和购物空间的同时，利用屋顶花园为倒班人员创造更多的休闲活动空间，并最大限度地将场地南侧优美的自然景观资源引入到场地之中，提升室内与庭院的空间品质；上部的轮班宿舍则采用合院式布局围合出完整的体量，其中南北侧为单廊南向宿舍布局，东西侧为单廊双向布局，以保证每间宿舍均可采到阳光；同时，将靠近机场高架一侧的房间处理为封闭阳台，以积极回应场地高速路噪音的影响。立面材料选用浅灰色涂料搭配金属百叶，简洁利落、美观节能并易于维护。整个设计期望打造人与自然交融、建筑与自然交融、城市与建筑交融的，具有生态、创新、便捷的生活服务设施综合体。

With the "Green·Blending" design concept proposed for the project, it aims to provide the users with a relaxing and convenient composite space environment. In the design, it tries to reorganize the functions of shift dormitories, living service and staff canteen, making vertical partitioning by means of opening and closure. The open street concept is introduced into the living service center and staff canteen at the lower part, and it can provide a convenient dining space for 6000 airport staff. Meanwhile, the roof garden is used to create more leisure activity space for the shift staff; the beautiful natural landscape resources on the south side of the site is introduced, to the maximum extent, into the plot to improve the interior and courtyard space quality. The shift dormitories on the upper part enclose into an integral mass by means of courtyard layout, of which single-corridor southward layout is applied for dormitories on the south and north sides, while single-corridor opposite-oriented layout is applied for dormitories on the east and west sides in order to ensure sunlight for each dormitory. At the same time, closed balconies are furnished for rooms close to the airport flyover to mitigate the impact of noise from the expressway. For the facade materials, light gray painting plus metal louvres are used as they are simple, beautiful, energy-efficient, and easy to maintain. It aims to create an ecological, innovative, and convenient living service facilities complex that blends people with nature, buildings with nature, and city with buildings.

首层平面 / First Floor Plan
标准层平面 / Typical Floor Plan

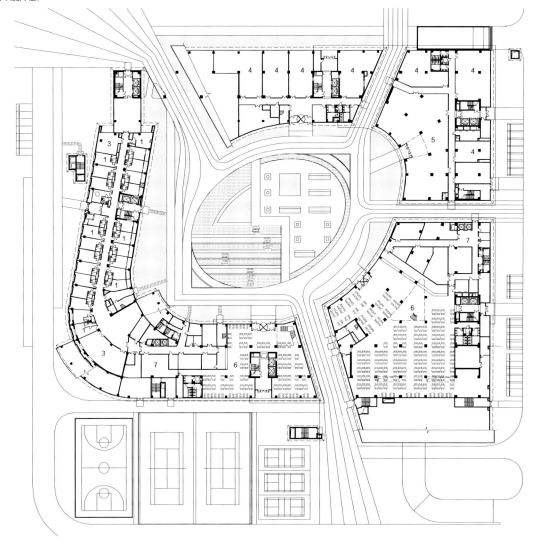

1 员工宿舍
2 管理人员宿舍
3 活动室
4 生活服务中心
5 咖啡厅
6 餐厅
7 厨房
8 休闲区
9 室外平台

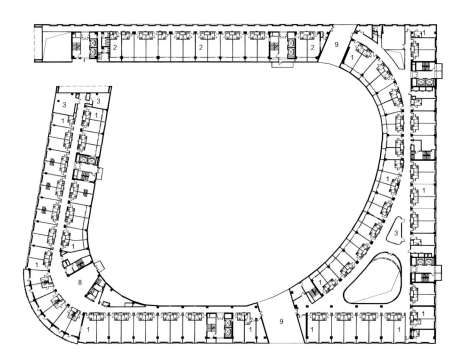

模型 / Models

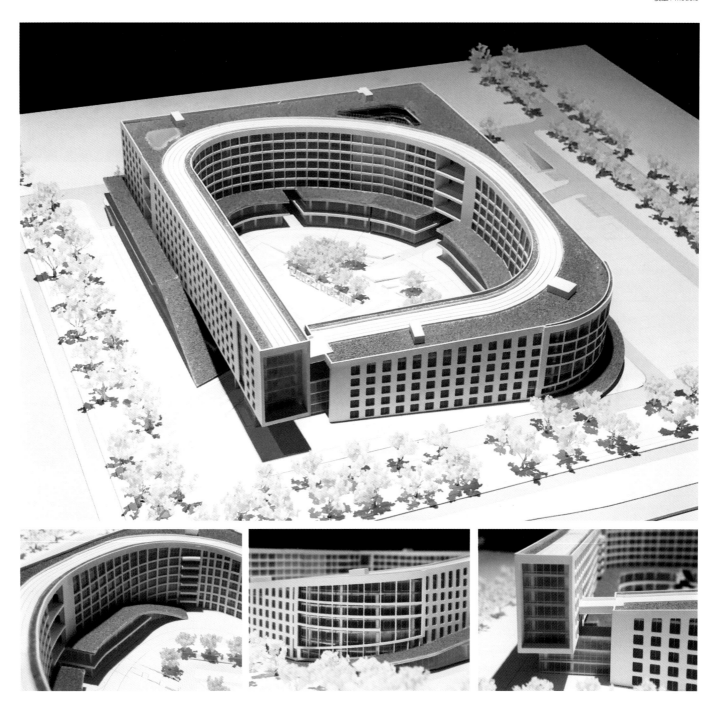

111

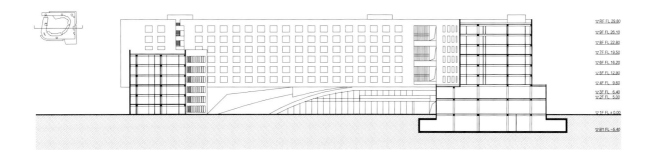

剖面 / Section

总平面 / General Plan

非主基地航空公司办公及宿舍项目
Non-main-base Airline Office & Dormitory Project
2017 ~ 2019, Beijing , 259703 ㎡

非主基地航空公司办公及宿舍项目位于北京新机场航站楼前核心区西北部，入场高架路以东的特殊位置上。在整体规划中，通过"共享能量树"将各航空公司办公楼有机地链接起来，兼顾各个航空公司独立性与建筑整体性的要求，达到功能相对独立、群组形象完整统一的有机整体。设计积极回应场域环境，采用综合办公围合内部公寓以及共享功能轴的"C"型结构布局。对外，保持完整的展示界面以彰显各航空公司的企业形象；对内，打开场地东侧界面，将东侧已规划的机场区中心景观带引入到基地W，最大限度利用周边自然景观资源，为使用者提供一个放松、惬意的复合型空间环境；机组公寓与出勤楼一端呈枝状分开，并做折窗处理，实现公寓间间有阳光、户户可观景的功能诉求；公共服务空间以开放、共享和高效原则集中设置。通过将综合办公楼、机组公寓与出勤楼，以及公共服务与优质景观有机地融合在一起，联系便捷，从而打造出既能让非主各航空公司职员惬意办公生活，又能进行对外形象展示的航空公司综合体。

The non-main-base airline office and dormitory project is located in the northwest part of the core zone in front of the terminal of Beijing New Airport, at the special position of east of the entrance flyover. In the master plan, the office buildings of different airlines are organically linked through "shared energy tree", meeting the requirements of independence of every airline and the architectural integrity, making it an organic whole with relatively independent functions and unified group image. The design actively responds to the field environment by adopting the C-shaped structure layout, in which the integrated office building encloses with the internal apartment building and a shared function axis is utilized. Outwards, a complete display interface is retained to manifest the corporate image of each airline. Inwards, the interface on the east side of the site is opened to bring in the central landscape belt in the planned airport zone on the east side, taking advantage of the ambient natural landscape resources to the maximum extent to provide the users with a relaxing and cozy composite space environment. The crew apartment and attendance building end are separate in dendritic manner, and folding casement is applied to meet the functional appeal of having access to sunlight and a view in each room of the apartment. Public service space is centralized in the principle of opening, sharing and efficiency. By organically merging the integrated office building, crew apartment and attendance building, public service and high-quality landscape with easy and efficient connection, it aims to deliver an airline complex that enables the staff of every non-main-base airline to enjoy their office life and every resident airline to display its image to the public.

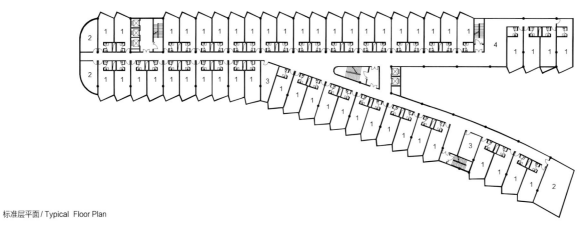

标准层平面 / Typical Floor Plan

1 宿舍
2 活动室
3 洗衣房
4 公共休息区

提起机场，人们马上想到的是"城市的门户""远行的起点和终点""迎来送往""悲欢离合"等等，毫无疑问，城市的机场已是现代城市生活越来越重要的节点，而节点中的航站楼和交通枢纽是主角，与之配套的导航塔台作为机场的核心，自然成为一个绕不开的建筑类型，往往会成为机场的标志物而备受关注。

对于塔台而言，"蔓设计"理念主要体现在基于机场总体环境，尤其是协调与航站楼关系的塔台形态、空间和功能的塑造，以及对于特定地域场所、文化、情感所形成的精神和心理意义的诠释，使塔台"介入"城市后能够对当下和未来产生可持续地积极影响。在满足视线无障碍的前提下，采用先进的集约型导航明室型制，能够降低塔台的造价成本。各项工程自建成启用以来，极大地提高了城市空管保障能力，较好地满足了航班量快速增长的需求，为飞行安全、顺畅提供了强有力的保障，并均已成为各大机场乃至所在城市的重要标志性建筑。

Whenever airport is mentioned, it would instantly remind people of "gateway of the city", "origin and destination of a long journey", "welcome and farewell", "partings and reunions", etc.; undoubtedly, urban airports have become an increasingly important node for modern urban life. Terminals and transportation hubs are the leading roles in the node, and naturally the supporting navigation control towers, as the kernel of airports, become an architectural type that cannot be bypassed; they receive significant attention as they would become the landmarks of airports in most cases.

For control towers, the concept of "organically-permeated design" is primarily reflected in the shaping of the form, space and function of control towers based on the overall environment of airports, especially when considering that control towers coordinate their relations with terminals, and in the interpretation of the spiritual and psychological significance formed about designated region and site, culture, and emotion; so that the control towers will exert sustained positive influence on the city for the present and future when they "step in". It will reduce the cost of control towers by adopting advanced intensive navigation bright rooms on the precondition of a barrier-free sight. Since they are put into operation these control towers have significantly improved the air control support capacity of the cities, meeting the needs of fast-growing flight volume, and providing strong guarantee for safe and smooth flights. They have even become the iconic buildings of the airports, or even their home cities.

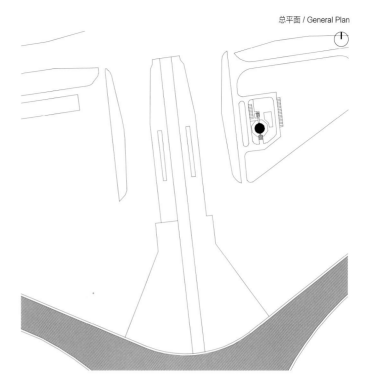
总平面 / General Plan

首都国际机场东区塔台
Eastern Air Traffic Control Tower of Beijing Capital International Airport
2004 ~ 2007 , Beijing , 2756 ㎡

首都机场东区塔台工程位于 T3 航管楼北侧 3 公里跑道的结束端，是首都机场 T3 区域的视线焦点。塔台的构思源于对先期确定的、由两个 Y 字对接在一起的 T3 航站楼方案的研究。将航站楼平面立体化形成了塔台基本形态，形成投影关系，自然地使两个尺度悬殊的建筑取得了呼应，而这一形态也真实地反映了现代导航塔台的功能本质，外装材料除塔身拟采用环氧型外墙涂料外，其余部位为复合铝板幕墙。明室部分因功能需要为落地玻璃窗（双层玻璃内夹电阻丝，可有效消除冷凝现象）。材料的搭配充分考虑了整个建筑的虚实对比、质感对比以及维护、清洗等方面因素。竖向的金属立挺使塔台向上的张力得到加强，弥补了 82 米限高带来的遗憾。

The project of the control tower for the east zone of Capital Airport is located on the finish end of the runway 3 km from the T3 air traffic control building, and it is the focus of attention in the T3 zone of Capital Airport. The conception of the control tower originates from the study of the previously-finalized T3 terminal scheme that resembles the shape of two "Y"s joined together. The basic shape of the control tower results from erecting the plan of the terminal, which makes the two buildings incommensurable in dimension echo to each other. And the form also gives a true reflection of the functional essence of a modern navigational control tower. Aluminum sandwich panel curtain wall is applied as the exterior decoration material for the control tower except for the tower body, which is finished with epoxy exterior wall paint. Electronically-heated double-glazed ground window is used for the bright room so that condensation can be effectively removed as required by the function of the room. Full consideration is given to the virtual-real contrast and textual comparison in the overall architectural appearance in selecting the exterior finish materials, and factors such as easy to maintain and clean are also taken into consideration. The straight vertical metal reinforces the upward strength of the tower to compensate the 82 meters height limit.

设计草图 / Sketch
剖面 / Section

平面 / First Floor Plan

0　　5m

1 门厅
2 消防值班
3 会议室
4 班前准备
5 讲评室
6 教室
7 管制员休班室
8 健身房
9 UPS
10 气体消防
11 直流配电
12 空调机房
13 小会议室
14 资料室
15 塔台主任办公室
16 副主任办公室
17 技术监控室
18 管制员休息室
19 明室检修环
20 塔台明室

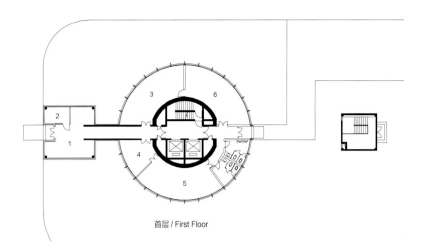

首层 / First Floor

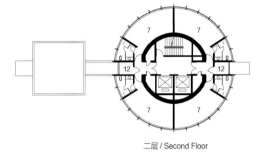

二层 / Second Floor

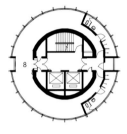

标高 +9.60 / Level +9.60

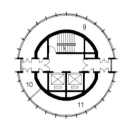

设备层 / Equipments Level

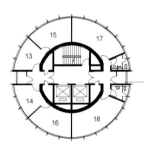

辅助管制层 / Auxiliary Control Level

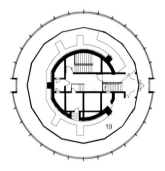

检修环层 / Repair Ring Level

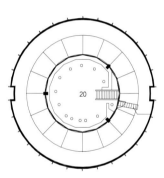

塔台明室 / Well-lighted Room

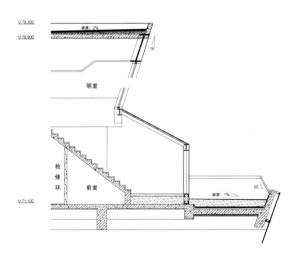

明室详图 / Details

总平面 / General Plan

西安咸阳国际机场新塔台及附属建筑
The New Air Traffic Control Tower of Xianyang International Airport and Its Affiliated Building in Xi'an
2009 ~ 2012 , Xi'an , 5600 ㎡

西安咸阳国际机场新塔台及附属建筑是西安咸阳国际机场二期空管工程的配套工程。方案秉承整体性、功能性、经济性和先进性原则,注重彰显"古城西安"的传统内涵,设计在满足空中航管功能要求的前提下,将这一极具现代化功能的建筑赋予更多的文化和精神内涵。塔台建筑造型取自西安城大、小雁塔的层叠、收分渐变的手法,形体玲珑秀丽、轮廓舒畅、比例优美;密檐式的百叶与玻璃幕墙的结合加深了建筑中的人文气息。塔身由实至虚,明室与检修环部位破茧而出,其造型语言结合现代机场导航塔台的功能特点来塑造,显得顺理成章;裙房的造型采用横向舒展的流线设计,肌理由密至疏,强化了塔台的高耸。统一的建筑手法与多方位的细部处理,恰如其分地呼应了机场高效、流畅的风格诉求,使整个建筑简约而不简单,精致优雅而不琐碎。

The new air traffic control tower of Xianyang International Airport and its affiliated building in Xi'an is a supporting project for the air traffic control project of Xia'an Xianyang International Airport Phase II. The design scheme adopts the principles of integrity, functionality, economy and advancement, paying attention on manifesting the traditional implication of Ancient Xi'an. On the precondition of meeting the requirements of flight navigation management function, the design endows more cultural and spiritual connotations to this building with extremely modern function. The architectural form of the control tower derives from the cascade and cone shape of the Big Wild Goose Pagoda and Small Wild Goose Pagoda; and the tower is exquisite and elegant, with a comfortable profile and delightful proportion. The combination of multi-eave louver and glass curtain wall enhances the humanistic air in the structure. With the tower body changing from the solid to the void and the component of the bright room and inspection ring breaking out of the cocoon, it is only natural to combine the functional features of a navigation control tower for a modern airport to create the shape of the control tower. Horizontally extended streamline design is adopted for the shape of the podium, with the texture changing from dense to sparse, so as to highlight the loftiness of the control tower. Unified architectural techniques as well as the multi-dimensional detail treatment do justice to respond to the appeal of an airport for efficiency and swiftness; as a result, the whole structure looks simple but not simplistic, delicate but not desipient.

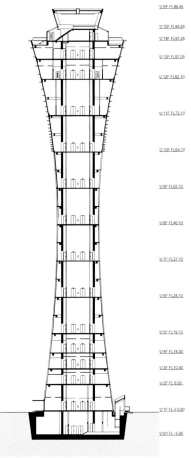

剖面 /Section

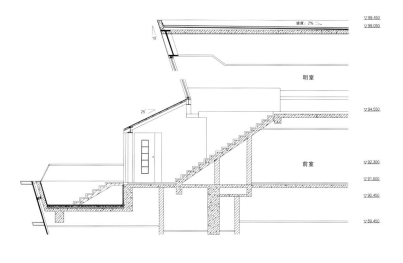

明室详图 / Detail

首层平面 / First Floor Plan
二层平面 / Second Floor Plan
四层平面 / Third Floor Plan
塔台平面 / Plan

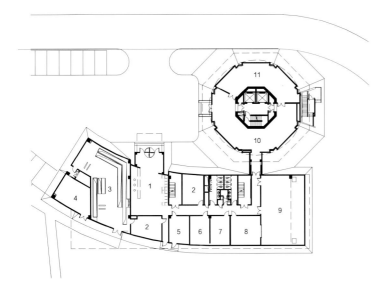

1 门厅
2 值班、更衣室
3 变配电室
4 柴油机房
5 消防控制室
6 气瓶间
7 弱电蓄电池
8 监控
9 通信机房
10 展廊
11 UPS间

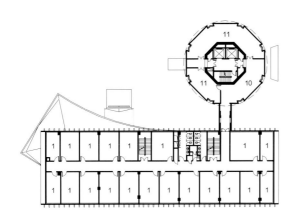
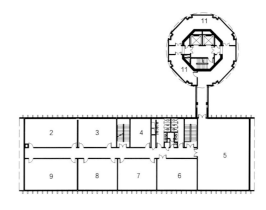

1 倒班宿舍
2 观测值班室
3 本站预报室
4 设备监控室
5 场监雷达机房
6 场监雷达监控室
7 观测总控
8 视频会商室
9 区域预报室
10 走廊
11 办公区

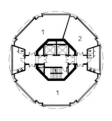
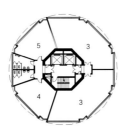
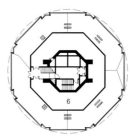

设备层 / Device Level   休息层 / Resting Level   检修环层 / Repair Ring Level   塔台控制室层 / Tower Control Room Level

1 更衣室
2 气瓶间
3 值班间
4 休息室
5 岗前准备室
6 检修环
7 明室

西安大雁塔　　　　　　　　西安小雁塔

设计意向 / Intentions of Design

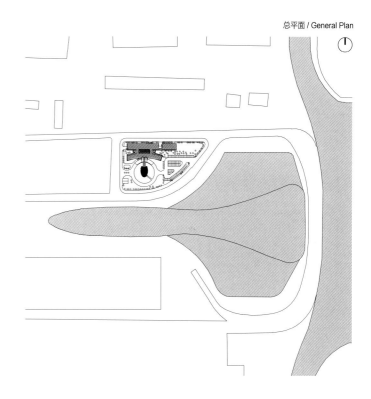
总平面 / General Plan

郑州新郑国际机场新塔台及附属建筑
The New Air Traffic Control Tower of Xinzheng International Airport and Its Affiliated Buildings in Zhengzhou
2013~2016, Zhengzhou, 13233 ㎡

郑州新郑国际机场地处中原腹地，是我国最大的重要干线机场及空中交通枢纽之一。项目位于郑州新郑国际机场综合交通换乘中心（GTC）交通枢纽北侧，包含塔台、航管楼两部分。方案注重从宏观环境分析项目要素，并结合地域文化元素，造型取自中原典型文化意向中的"贾湖骨笛"（现为河南省博物馆的镇馆之宝）与商代青铜酒器"觚"的形象，以谦逊的弧形姿态伫立在机场轴线北侧，以烘托和强调机场主轴线及航站楼的主角地位，并与航站楼的流线型设计相得益彰，相辅相成，以提高机场总体环境的整体性和协调性。新郑塔台是我国首座最高的非线性清水混凝土建筑，高93.5米。在满足使用功能要求下，采用了国际领先的建筑型制、经济高效的结构选型及高集成化的施工工艺。项目在方案创作、施工图设计、设计配合施工等全阶段中应用BIM技术，保障了设计完成度和施工精度，并使后续使用、维护及管理更加高效。建成后的塔台以和谐的建筑形态融入城市空间，以谦逊之姿、欢迎之姿迎来送往，全面展示了郑州大气、庄重、古韵的城市特质，彰显出"中原崛起"的时代精神。

Zhengzhou Xinzheng International Airport is located in the hinterland of Central Plains, and it is one of China's biggest trunk line airports and air traffic hubs. This project is on the north of the Zhengzhou Xinzheng International Airport GTC, including a control tower and an air traffic control building. The design scheme analyses the project factors from macro-environment, combined with the regional cultural elements. The shape of the control tower originates from the "Jiahu bone flute" (it is currently the centerpiece of Henan Museum), a typical cultural image of the Central Plains and the image of "Gu", a kind of bronze drinking vessel popular in the Shang dynasty (ca. 1600 B.C. ~ ca. 1046 B.C.). The control tower stands on the north side of the airport axis with its humble arc-shaped posture, so as to foster and emphasize the leading role position of the airport axis and terminal, benefitting a company with the streamlined design of the terminal; hence, the integrity and coherence of the overall environment of the airport is uplifted.

The Xinzheng control tower is China's first and highest non-linear fair-faced concrete structure, with the apex of 93.5m. While the requirement of utility function is met, what is applied in the design and construction processes are internationally advanced architectural type, economically efficient structure selection, as well as highly integrated construction technology. BIM technique is applied throughout the design process, from conceptual design to construction documents design, and from the design coordination to construction, so that the design schedule and construction accuracy are guaranteed; and subsequent utilization, maintenance and management are more efficient. The as-built control tower blends in with the urban space with its harmonious architectural shape, greeting and bidding farewell with the humble and welcoming posture; it displays Zhengzhou's urban traits as being grand, solemn and archaic, manifesting the ethos of the time during the "rise of the Central Plains".

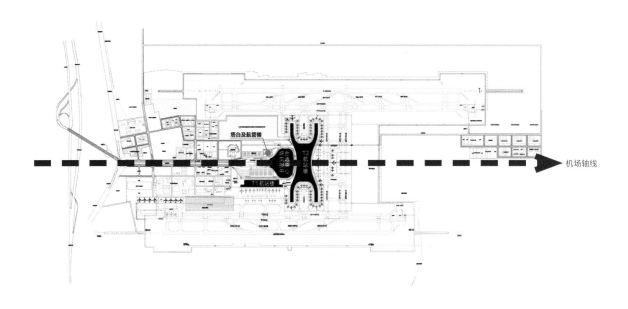

轴线分析 / Axis Analysis

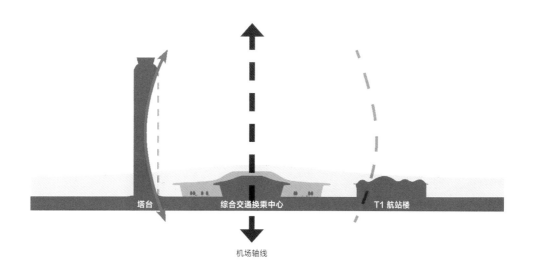

（1）城市空间的诠释——谦逊之姿、欢迎之姿
Interpretation of the urban space —— a gesture of modesty and welcome

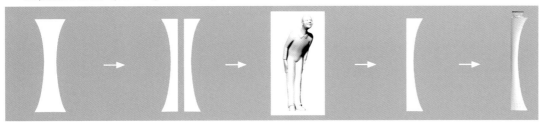

（2）塔造型的文化意象之一——"贾湖骨笛"
The first contextual heritage in the form of the control tower —— "Jiahu bone flute"

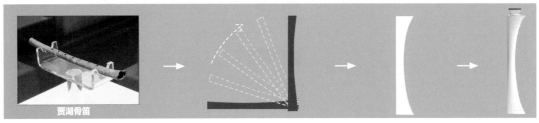

（3）塔造型的文化意象之二——"觚"
The second contextual heritage in the form of the control tower —— "Gu"

概念生成 / Concept Generation
模型 / Model

首层平面 / First Floor Plan
三层平面 / Third Floor Plan
塔台平面 / Plan

0 5m

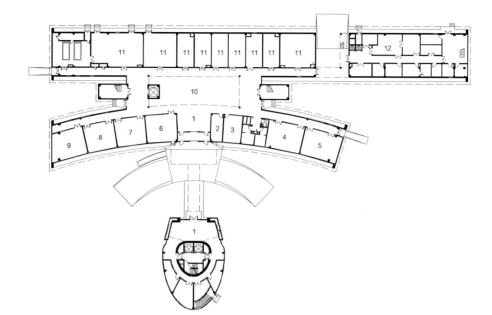

1 门厅
2 值班室
3 复印室
4 飞行服务
5 办公室
6 消防控制室
7 会议室
8 档案室
9 技术资料室
10 中庭
11 设备机房
12 厨房

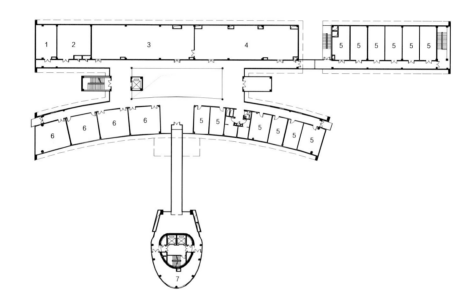

1 设备器材库
2 设备维修间
3 运行监控室
4 设备机房
5 办公室
6 休息室
7 备用室

标高 +13.80 / Level +13.80

标高 +71.70 / Level +71.70

标高 +75.60 / Level+ 75.60

标高 +79.50 / Level +79.50

标高 +84.50 / Level +84.50

标高 +87.70 / Level +87.70

1 休息室
2 储备间
3 气象监测室
4 备用间
5 明室

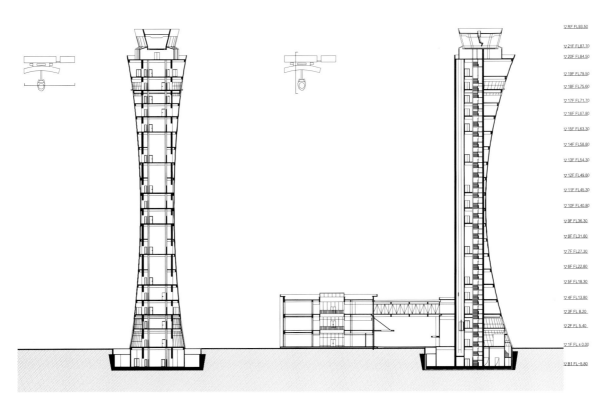

剖面 / Section

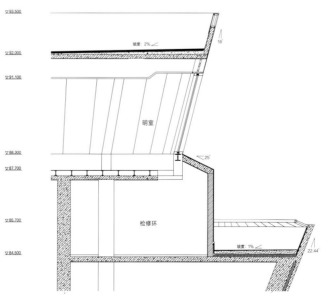

明室详图 / Detail

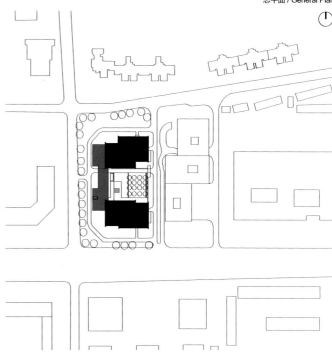

总平面 / General Plan

北京按摩医院

Beijing Massage Hospital
2016 ~ 2019, Beijing, 38995 ㎡

北京按摩医院新址位于北京市朝阳区广渠路 36 号地块，区位优越、交通便利。新院的建设将全面致力于满足和适应医院未来需求，成为我国面向世界弘扬中医文化和对弱势群体关怀的人权事业的重要展示窗口。面对这一人文历史背景悠久且具有特殊意义的医疗建筑，需要打破传统医院设计模式，突出以人为本的核心价值，并引入和贯彻阳光医疗、景观医疗、绿色医疗的最高设计原则。设计提炼中国传统建筑的"檐""梁""柱"经典元素，将其转译并抽象成新的建筑语汇，与建筑功能有机融合在一起，建筑细部点缀应用了斗栱、月亮门、屏风、漏窗等传统手法；同时，建筑色彩采用北京传统建筑中常见的深灰、浅灰色与原木色作为主要建筑用色，体现对于整个城市肌理的传承与延续，最大限度的唤起按摩医院特有的地域与场所精神。设计将抽象后的中国传统建筑要素重塑与整合，使形式追随功能，实现有机建构，从而打造真正体现人文关怀的、现代且又具有中国特色的医疗建筑作品。

The new Beijing Massage Hospital is situated at Plot 36 of Guangqu Rd., Chaoyang District, Beijing, boasting a superior location with convenient transportation. The construction of the new hospital will be fully committed to accommodating the future demand of the hospital, and becoming an important exhibition window for China to promote traditional Chinese medical culture across the world and to reinforce the cause of human rights through caring the social vulnerable groups. For the design of the medical buildings with a long historical background and special significance, it is necessary to break the design pattern for traditional hospitals, to highlight the core values of people-orientation, and to introduce and apply the supreme design principle of sunlight medical treatment, landscape medical treatment, and green medical treatment. Classical elements in traditional Chinese architecture, such as eave, girder and column are extracted and translated into new architectural vocabulary, which are organically integrated with the architectural functions: traditional components like bracket sets, moon gate, screen and lattice window are applied here and there in the architectural details. Meanwhile, in terms of architectural colors, dark gray, light gray and burlywood, which are frequently used in traditional buildings in Beijing, are applied as the key architectural colors to represent inheritance and continuance of the whole urban context, and to arouse, to the maximum extent, the spirit of region and place that is unique to the massage hospital. The design remolds and integrates the abstracted traditional Chinese architectural elements, to subject forms to functions to achieve organic construction so that a medical architectural work is created that truly demonstrates humanistic care, modern traits and Chinese characteristics.

首层平面 / First Floor Plan
二层平面 / Second Floor Plan

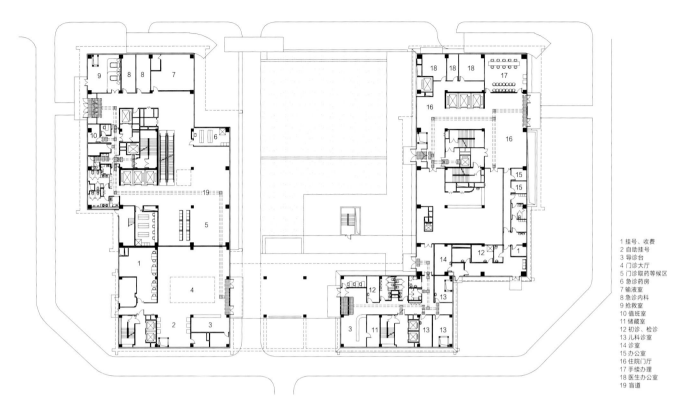

1 挂号、收费
2 自助挂号
3 导诊台
4 门诊大厅
5 门诊取药等候区
6 急诊药房
7 输液室
8 急诊内科
9 抢救室
10 值班室
11 储藏室
12 初诊、检诊
13 儿科诊室
14 诊室
15 办公室
16 住院门厅
17 手续办理
18 医生办公室
19 盲道

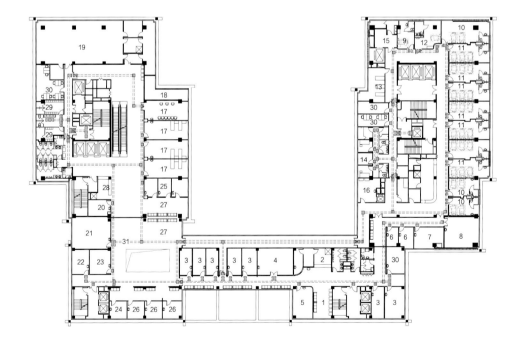

1 挂号、收费
2 导巡台
3 儿科诊室
4 针灸室
5 针刀室
6 ST 理疗室
7 OT 理疗室
8 PT 理疗室
9 单人病房
10 双人病房
11 三人病房
12 无障碍病房
13 按摩大厅
14 值班室
15 污物间
16 配膳间
17 治未病诊室
18 采血间
19 中心检验室
20 耳鼻喉科诊室
21 口腔科诊室
22 皮肤科诊室
23 眼科诊室
24 骨伤科诊室
25 内科诊室
26 外科诊室
27 候诊大厅
28 护士站
29 更衣室
30 办公室
31 盲道

标准层平面 / Typical Floor Plan
模型 / Models

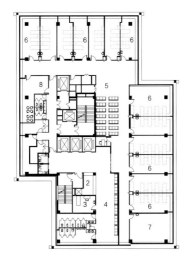

1 收费
2 护士站
3 初诊
4 一次候诊
5 二次候诊
6 普通按摩诊室
7 中医治疗室
8 示教室
9 病房
  （双人／三人／六人）
10 无障碍病房
11 按摩大厅
12 办公室
13 值班室
14 配膳间
15 污物间

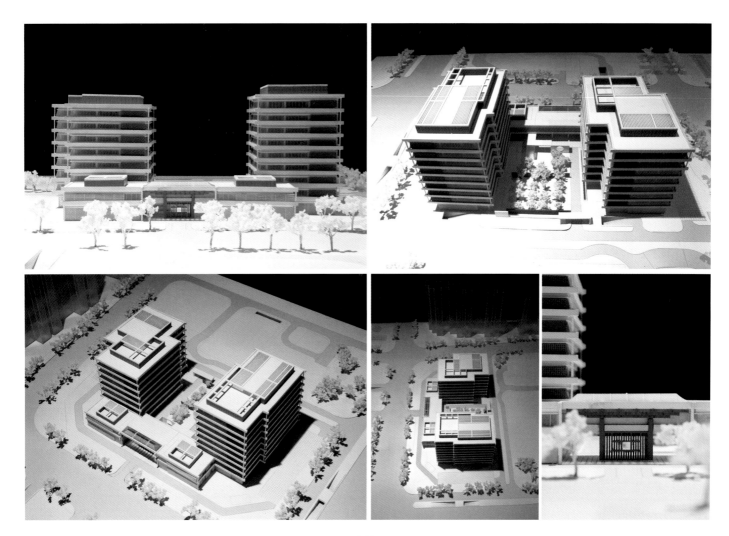

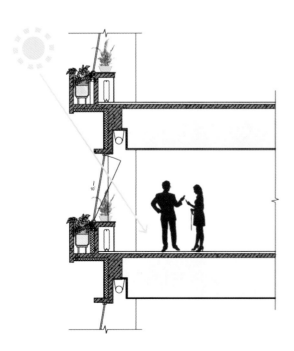

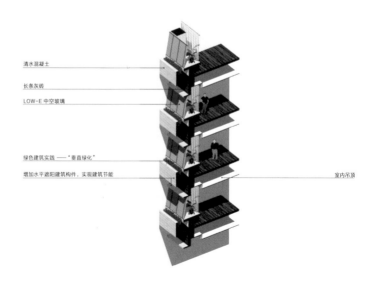

清水混凝土

长条灰砖

LOW-E 中空玻璃

绿色建筑实践——"垂直绿化"

增加水平遮阳建筑构件,实现建筑节能

室内吊顶

绿色技术 / Green technology

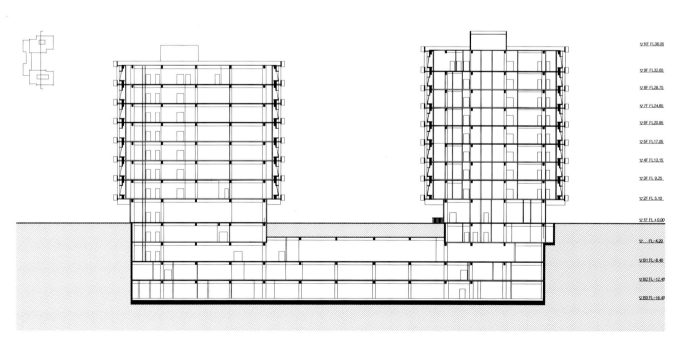

剖面 / Section

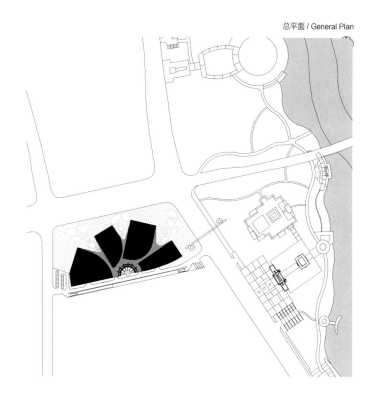

总平面 / General Plan

阿克苏三馆及市民服务中心

Aksu Public Complex and Civic Service Center
2015 , Xinjiang , 59800 ㎡

阿克苏三馆及市民服务中心项目是集博物馆、文化馆（含美术馆）、图书馆以及市民服务功能为一体的文化综合体，位于阿克苏市新区的多浪河西岸。设计提出"多元汇聚"的概念，将代表着历史、知识和艺术的博物馆、图书馆和美术馆，以及体现着人本关怀的市民中心，汇聚到象征着"大美和顽强"精神的胡杨林下，组成了盛开的雪莲花。从城市到场地、从理性到浪漫、从规整的几何体到非线性的空间和体量的推敲，最终融合形成一个有机的整体。按各个建筑的功能特点组织的动线系统充分考虑了现代展示技术的应用，以期为观众带来全新的使用体验。丰富的形态结合了取自阿克苏白水河里的水刷石材料，建筑立面谨慎的开窗处理充分考量了新疆维吾尔自治区阿克苏地区干热气候的特点，也使建筑显得内敛、神秘而又浪漫。

Situated on the west bank of Duolang River in the new district of Aksu, Aksu public complex and civic service center project is a cultural complex that integrates the function of museum, cultural center (including gallery) and library, as well as civic service. The design proposes the concept of "multielement convergence", gathering a museum, library and gallery representing history, knowledge and art respectively, as well as a human-oriented civic center under a populus euphratica forest, a token of supreme beauty and indomitable spirit, the group of buildings resemble a snow saussurea in blossom. Pondering from city to place, from reason to romance, from well-shaped geometry to nonlinear space and mass, all these elements are ultimately integrated into an organic whole. The circulation system organized in light of the function feature of each building gives full consideration of the application of modern exhibition technologies in a bid to bring new user experience for visitors. The variety of forms also contains the granitic plaster originated from the Baishuihe River in Aksu. The discreet windowing on the facade takes into full consideration the dry hot climate of Aksu, Xinjiang, which also makes the building complex appear to be reserved, mysterious and romantic.

STEP 1: 基地分析——分析基地所处位置与周边情况。

Site analysis —— to analyze the location of the site and its ambient circumstances.

STEP 2: 聚集——人 / 市民中心，知识 / 图书馆、历史 / 博物馆、艺术 / 美术馆，在用地内交融。

Aggregation —— people/citizen center, knowledge/library, history/museum, and art/gallery are integrated in the lot of land.

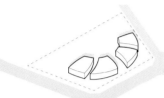 

STEP 3: 基本体量——在场地肌理上生成建筑形体。

Basic mass —— to generate the architectural shape on the site texture.

STEP 4: 开敞——沿城市界面敞开，形成开放与聚集的城市景观。

Openness —— open to the urban interface to form a cityscape mixing openness and aggregation.

STEP 5: 广场——聚集式布局使每个场馆具有各自独立的广场空间。

Plaza —— aggregated layout enables each building to have its independent plaza space.

STEP 6: 整合——置入连廊、塔，使得三馆一中心既统一又相互独立。

Completion —— placing corridors and a tower enables the museum, library, gallery and citizen center both unified and separate.

概念生成 / Concept Generation
新疆印象 / Impression of Xinjiang

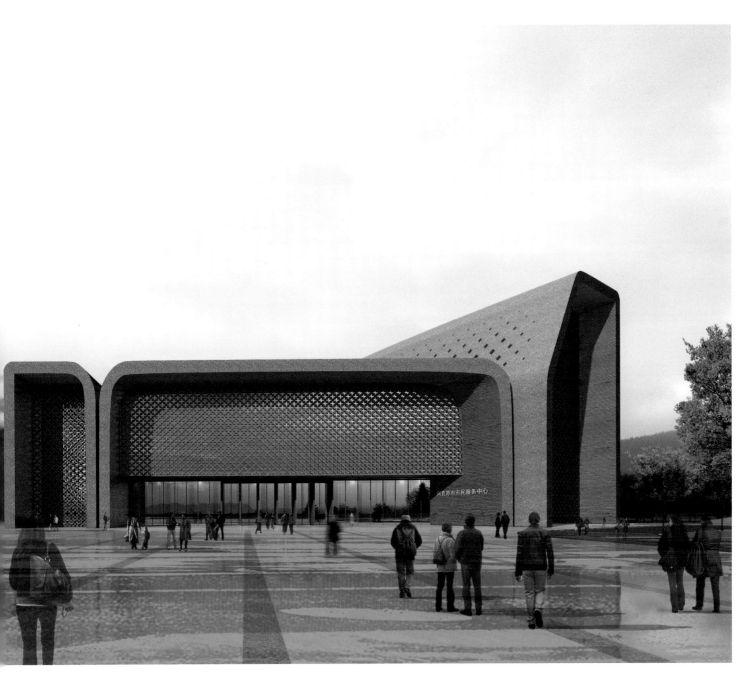

首层平面 / First Floor Plan

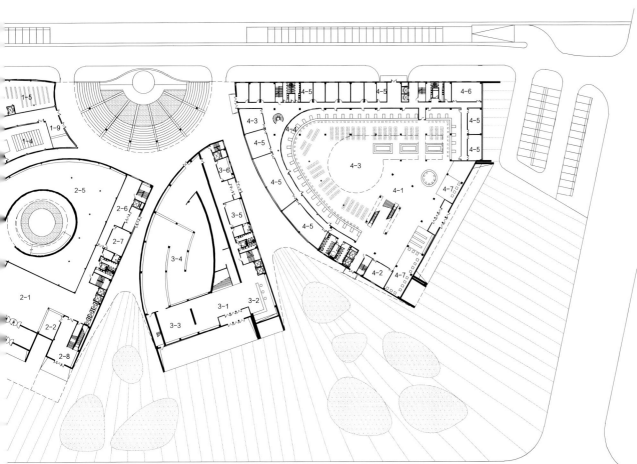

1 图书馆
1-1 入口门厅
1-2 陈列展览厅
1-3 小报告厅
1-4 大报告厅
1-5 古籍书库
1-6 综合活动室
1-7 培训室
1-8 管理
1-9 小卖部

2 博物馆
2-1 入口门厅
2-2 服务台
2-3 票务
2-4 序厅
2-5 3D多媒体展廊
2-6 办公门厅
2-7 贵宾室
2-8 出口门厅

3 文化馆
3-1 入口门厅
3-2 接待区
3-3 序厅
3-4 美术馆展厅
3-5 VIP接待室
3-6 办公室

4 市民服务中心
4-1 行政服务大厅
4-2 值班、管理
4-3 等候、休息
4-4 窗口
4-5 办公室
4-6 会议室
4-7 商务中心

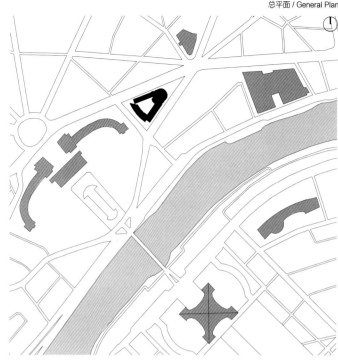

总平面 / General Plan

巴黎世界环保展示中心及企业孵化器
The World Environmental Protection Exhibition Center and Business Incubator in Paris
2017, Paris, 1600 ㎡

本项目的设计范围位于法国经济社会委员会大楼地下1、2层。法国经济社会委员会大楼坐落于巴黎第十六区,处于特洛卡迪罗广场轴线一侧,西南侧紧邻夏约宫和法国历史建筑博物馆,北侧是威尔逊总统大道,东南侧距离塞纳河仅200米,距离埃菲尔铁塔仅600米。法国经济委员会大楼始建于1925年,前后经历过两次加建,是世界上第一座加强混凝土建筑。业主需要对原有建筑的地下空间进行改造,并作为世界环保展示中心及企业孵化器的新址。面对地处特殊历史地段的改造工程,设计初始阶段依据改造强度的不同提出了两种改造策略:(1)自然之核——设计基于地下展示空间的展示动线,为原有建筑内院创造出一条自然路径,使其成为新旧建筑、地下与地上空间联系的介质,将阳光、景观、人、旧建筑引入地下空间,形式追随功能,实现有机建构;设计保留了原有建筑的结构体系,最大限度地减弱了对原有建筑的影响。(2)孵化器——新的功能需求是为新型的环保企业提供技术支持及展示空间,因此设计中在建筑内院植入了一个"胚胎",为地下空间提供充足自然光线的同时激活内部庭院空间,与建筑功能有机地融合在一起。"保护与更新"一直是围绕特殊历史地段改造项目的核心设计理念。面对设计中新的功能需求,本项目采取了着重强调"有机更新"的设计策略,建筑应随着功能需求的变化而发生变化,同时为其注入新的活力,使其与旧建筑有机地融合在一起。

The scope of design for this project is basement level 1 and level 2 of The French Economic, Social and Environmental Council (CESE). CESE is located in 16E ARR. Paris, on one side of the Trocadero Plaza axis, close to Palais de Chaillot and Cite de l'architecture et du patrimoine on the Southwest side; Avenue du President Wilson is to the North side, and it is 200 m from the Seine River, and 600 m from la Tour Eiffel. The CESE Building was built in 1925 and it is the first reinforced concrete building in the world after two extensions. The client needs to transform the underground space of the original building, and to make it the world environmental protection exhibition center and business incubator. In the face of a modification project situated on a special historic section, two modification tactics are proposed in light of different intensity of modification in the initial stage of design: (1) Kernel of nature —— based on the exhibition circulation in the basement exhibition space, the design creates a natural path for the inner court of the existing building to make it a medium to connect the new building with the old one, the underground space with the ground space, and the sunshine, landscape, people and the old building are thus introduced into the underground space; organic construction is achieved when form follows function. The design retains the structural system of the existing building so that impact on the old building is minimized. (2) Incubator —— the new functional requirement is to provide new environmental protection enterprises with technical support and exhibition space; therefore, the design implants an "embryo" into the inner yard, and it activates the inner yard space while providing the basement with sufficient natural light, organically blending in with the architectural function. Preservation and renewal has always been the core design concept over the transformation project at a section with a special history behind it. In the face of the new functional need in the design process, this project adopts the design tactic emphasizing "organic renewal". Architecture should follow the changes of functional needs; meanwhile, it injects new vitality into it so that it organically blends in with the old building.

设计策略 / Design Strategy

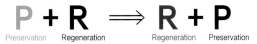

P + R ⟹ R + P
Preservation Regeneration   Regeneration Preservation

设计范围 / Design Area

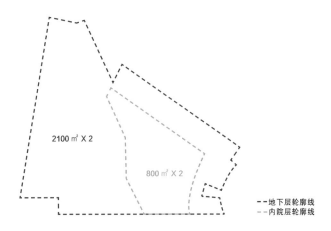

2100 ㎡ X 2
800 ㎡ X 2

— — 地下层轮廓线
— — 内院层轮廓线

现状功能分析 / Function Analysis

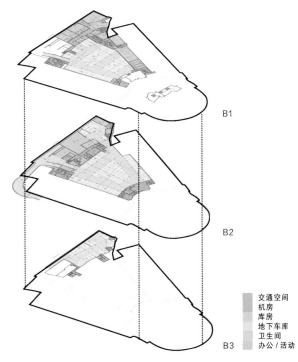

B1
B2
B3

交通空间
机房
库房
地下车库
卫生间
办公 / 活动

现状评价 / Status Evalution

现状实景 / Views

改造条件评价 / Situation of the C.E.S building

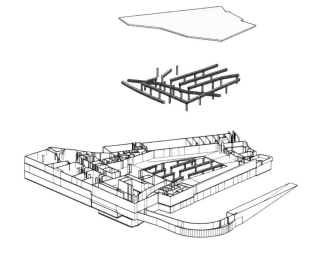

总平面 / General Plan

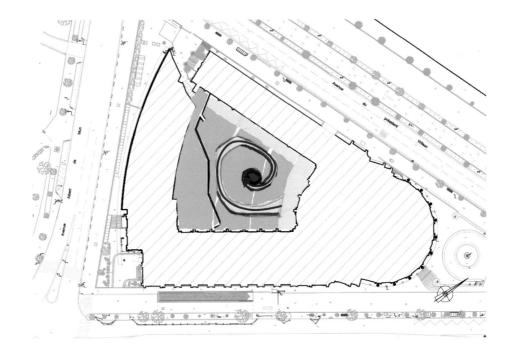

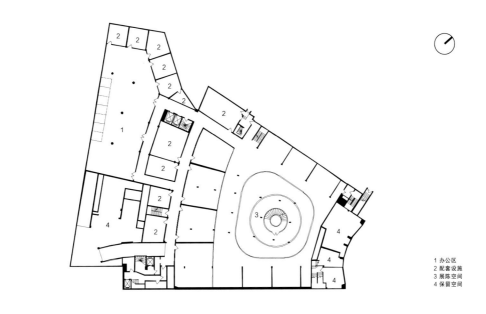

1 办公区
2 配套设施
3 展陈空间
4 保留空间

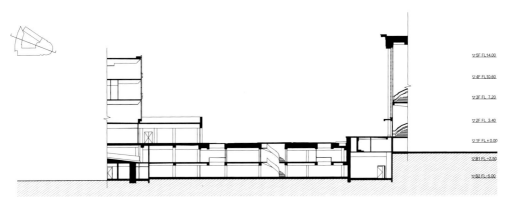

地下一层平面 / Ground Floor Plan
剖面 / Section

"自然之核"概念草图 / "Kernel of Nature" Concept Sketch
"孵化器"概念草图 / "Incubator" Concept Sketch

总平面 / General Plan

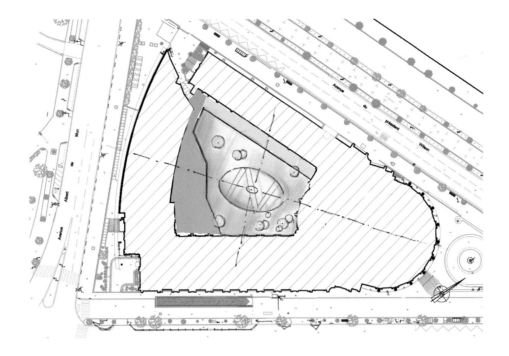

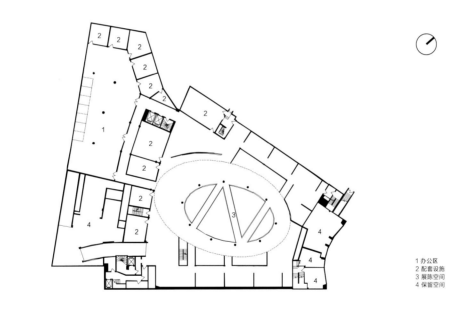

1 办公区
2 配套设施
3 展陈空间
4 保留空间

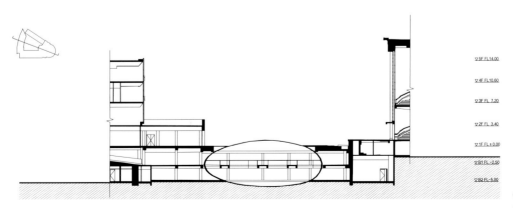

地下一层平面 / Ground Floor Plan
剖面 / Section

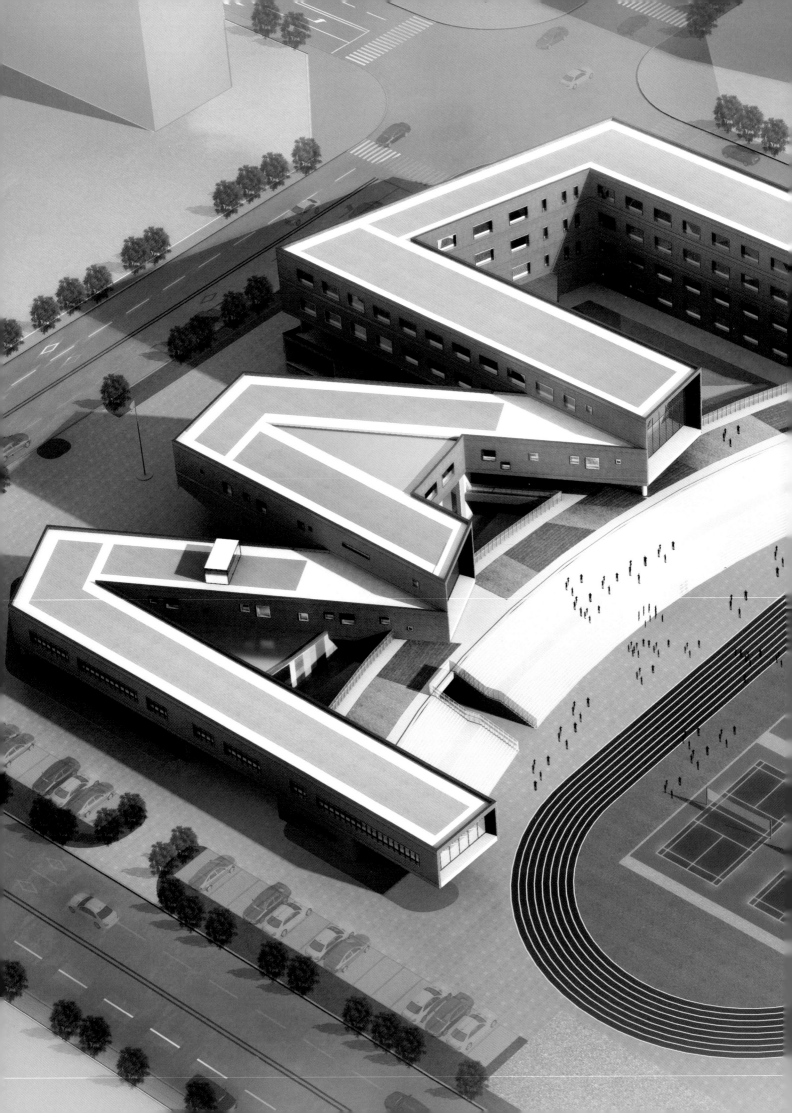

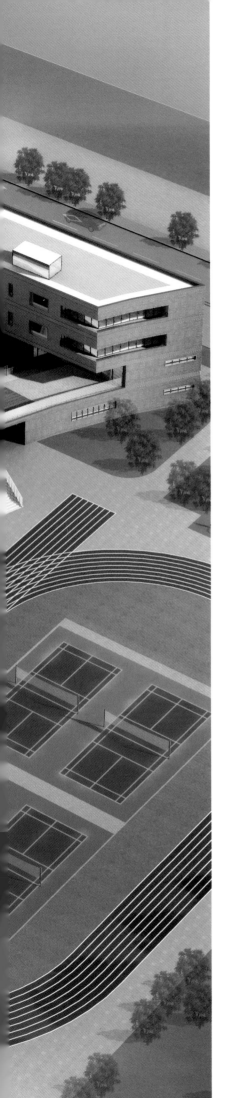
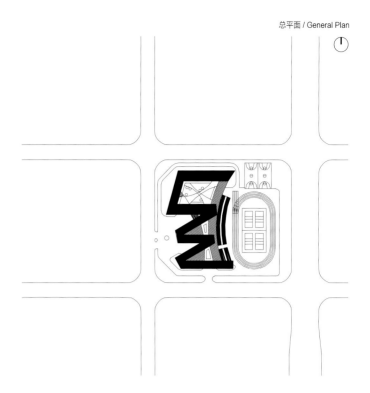

总平面 / General Plan

郑州航空港银河办事处第二邻里中心小学

Zhengzhou Airport Yinhe Office No.2 Neighborhood Central Primary School
2014, Zhengzhou, 13232 ㎡

小学是在孩子成长过程中塑造创造力极为重要阶段。郑州航空港银河办事处第二邻里中心小学位于河南郑州航空港经济综合实验区内，未来将成为周边社区生活中的人性化和活跃因素，在城市中完成教育功能的同时，体现出新型城市人文和友好的一面，促进整个社区生活的和谐与活力形成。设计的出发点就是在规矩有序的新区规划环境背景中，在尊重基地周边建筑尺度的前提下，探讨有机地构建符合青少年身心成长的充满趣味的空间序列。建筑功能按照普通教室、公共教室、专用教室等不同属性空间的分区布置，理性而高效；与此同时，建筑体量则呈现从规整到自由、从静态到动态、从封闭到开放、从严谨到活泼之间的转变，从而形成一个连续的、丰富的教学及游戏空间，并且之字形设计能够营造出更多的不同功能的室外活动空间。切合人性尺度而有趣的内外空间极好地对应了青少年的性格特点，让其在教学与游戏、室内与室外之间富有变化的空间体验过程中，激发珍贵的创造力和对学校生活的热情，在一个具有人文关怀的环境中健康成长。

Primary school education covers an extremely important span of time for shaping the children's creativity during their process of growth. Zhengzhou Airport Yinhe Office No.2 Neighborhood Central Primary School is located in Zhengzhou Airport Economic Comprehensive Experimental Zone, and it will become the focused and active element in the ambient community life. While it provides the urban educational function, the school also helps demonstrate the new-type urban humanistic and friendly aspects, promoting the harmony and vitality in the community life. The objective of design is to explore organic construction of the spatial sequence that is in line with the requirements of minors' physical and mental growth, and that is full of fun in the context of the orderly planning of the new zone and under the precondition of respecting the architectural scale of the ambient buildings. In the design space with different functional attributes, such as ordinary classrooms, public classrooms and special classrooms are placed in separate zones, which is both reasonable and efficient; meanwhile, the architectural space changes from the regular to the varied, from the static to the dynamic, from the closed to the open and from the rigid to the vivacious, therefore shaping a continuous, and varied teaching and game area; in addition, the z-shaped design enables to create more space with varied functions for outdoor activities. The indoor and outdoor space that is interesting and in compliance with the pupils' dimensions well responds to the minors' personality characteristics, enabling them to activate valuable creativity and develop enthusiasm for campus life during the process of experiencing the varied space for teaching and game, and for indoor and outdoor activities. It aims to facilitate the healthy growth of the pupils in an environment full of humanistic care.

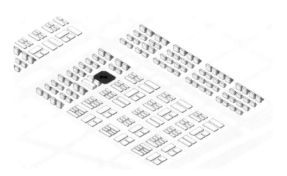

STEP 1：基地范围与场地关系。
The relationship between the site scope and the place.

STEP 2：相邻地块体量。
The mass of the neighboring lot.

STEP 3：基地辐射范围
The direction of radiant scope buildings.

STEP 6：建筑空间从规整到变化、静态到动态、封闭到开放、严谨到活泼。
The architectural space changes from the regular to the varied, from the static to the dynamic, from the closed to the open and from the rigid to the vivacious.

STEP 7：形成入口广场空间。
Forming a entrance plaza.

STEP 8：抬起建筑形体
The building is at the entrance.

概念生成 / Concept Generation

162

流方向与相邻建筑影响。
human circulation within the and the impact of ambient

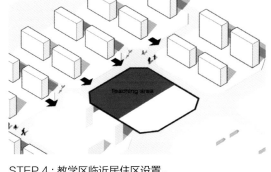
STEP 4：教学区临近居住区设置。
The teaching zone is placed near the residential zone.

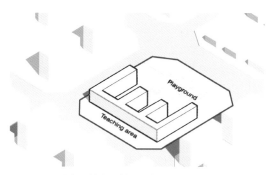
STEP 5：教学区基本建筑体量与操场布置。
The basic building mass in the teaching zone and the layout of the playground.

接送等待空间。
ate a waiting area for parents

STEP 9：专业教室与游戏空间串联起整个场地室内与室外、动态与静态空间。
The classrooms and game zone connect the interior and exterior as well as the static and dynamic space in the whole site.

STEP 10：赋予低碳节能材质，概念生成完成。
Adopting low-carbon energy-saving materials, and concept generation completed.

模型 / Models

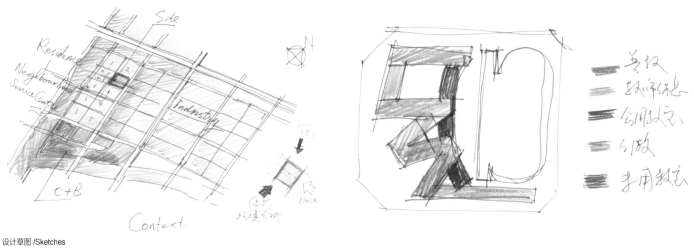

设计草图 /Sketches

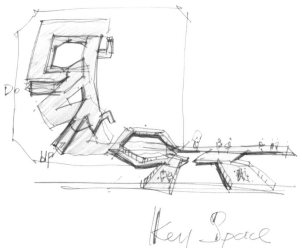

Key Space

首层平面 / First Floor Plan
二层平面 / Second Floor Plan

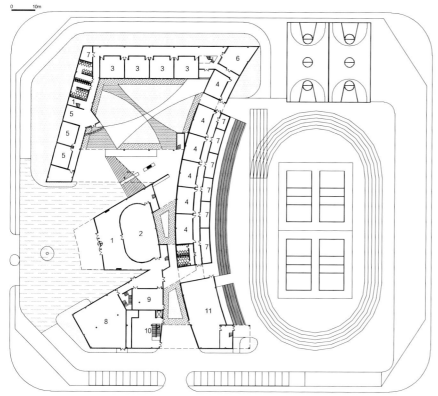
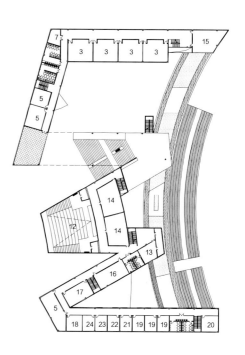

三层平面 / Third Floor Plan
四层平面 / Fourth Floor Plan

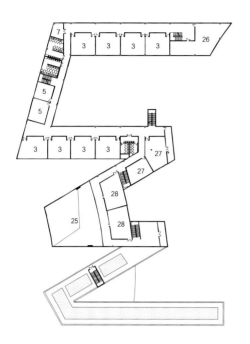
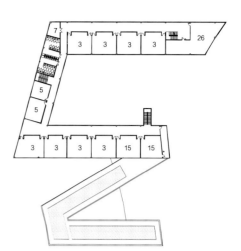

1 门厅 / 展厅
2 校史展室
3 普通教室
4 实验教室
5 教师办公室
6 总务及维修
7 仓库
8 学生餐厅
9 教师餐厅
10 备餐间
11 风雨操场
12 报告厅
13 德育展室
14 音体教室
15 活动室
16 心理咨询
17 体能测试
18 学生社团办公室
19 教务办公室
20 校长办公室
21 档案室
22 会议室
23 文印室
24 广播室
25 报告厅上空
26 阅览室
27 艺术教室
28 多媒体教室

总平面 / General Plan

北京雁栖湖山地艺术家工作区
Beijing Artist Studios on Yanqi Lakeshore
2016, Beijing, 119000 ㎡

本项目位于国际会都·北京雁栖湖畔，该区域是我国2014年APEC国际会议举办地，山林湖泊风景旖旎。基地山地特征明显，地块内最大高差达40米，中间贯穿两条城市水景绿地。设计以"自然·生长"为概念，意在以敬畏自然的姿态，让建筑顺应自然环境的指引而生长并取得有机的平衡。通过分析利用不同的坡度走势，形成建筑台地并提出对应的布局方案和观景策略，使建筑与富有挑战性、复杂多样的地势巧妙结合；采用人性化、创造性和因地制宜的设计手法，将"湖、山、河、林、谷"五个景观要素、地域特色文化、丰富的建筑空间和富有亲切感的地域性建筑材料巧妙而有机地融合，多方面提升建筑品质，最大限度地激发艺术家的创作灵感，使其充分享受远离城市喧嚣、回归自然、亲近自然的惬意和静谧，同时力求实现用地价值的最大化，使之成为具有建筑学意义的休闲办公建筑的样板作品。

The project is located on the shores of Yanqi Lake in Beijing, an international conference site, where APEC International Conference 2014 was held in China, boasting exquisite scenery of mountainous forest and lake. The site is with obvious features of a hilly area and the maximum altitude difference in the site is 40m, with two urban waterscape greenbelts running through the site. Based on the concept of "nature · growth", the design aims to take the attitude of revering nature and enable buildings to grow by complying to the guidance of natural environment and reach organic equilibrium. Through analyzing and employing various movements of the slopes to shape building mesas, relevant layout scenarios and landscape strategies are proposed so that buildings are artfully blended with the challenging and complex terrain. Humanistic, creative and flexible design approach is adopted to organically and ingeniously integrate the five landscape elements of "lake, hill, river, forest and valley" with regional culture, various architectural space and regional construction materials giving off a cordial feeling, in a bid to uplift the architectural quality from various dimensions, to stimulate the creative inspiration of artists to the maximum extent, so that they can fully enjoy the pleasure and tranquility of being isolated from the hustle and bustle of the city, returning to nature and being close to nature. Meanwhile, it aims to maximize the value of the land use, making it a model work of leisure and office building in the sense of architecture.

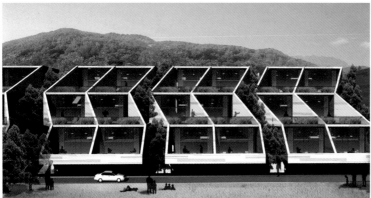

别墅效果 / Villa Views

分析图 /Analysis

等高线分析 / Contour Analysis

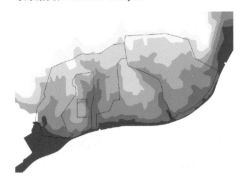

STEP 1：原始等高线
The original contour.

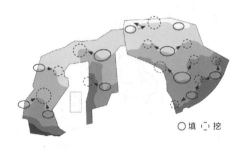

STEP 2：提炼等高线走势，土方平衡
Extract the contour movement tendency based on the original contour and achieve earthwork balance

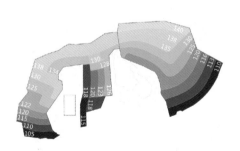

STEP 3：形成最终建筑台地，尽最大可能
保证每栋建筑观景
Shape the ultimate building mesas in order to assure the view of every building at the utmost.

退台式布局分析 / Stepped Form Analysis

景观与朝向 / Landscape and orientation

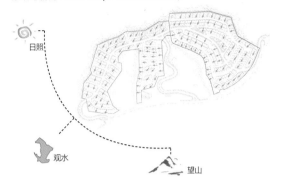

地段价值分析与评估 / Analysis and evaluation of the value

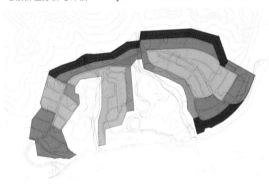

道路竖向设计 / Design of Vertical

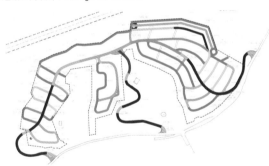

针对不同坡度选择对应的地形利用方式。若坡度≤25%，采用组团退台式布局，则上下两户之间，高差在3米左右；若坡度＞25%，采用每户退台的方式，则每户高差可达4米，从而保证其后排建筑单元的观景需要。

Pertinent approach of terrain utilization should be adopted for a different gradient. If the group set-back layout is adopted when the gradient is ≤ 25%, then the height difference between the two floors is around 3m. If the unit set-back layout is adopted when the gradient is > 25%, then the height difference between the two floors can reach 4m, so as to guarantee the view for the building blocks at the rear row.

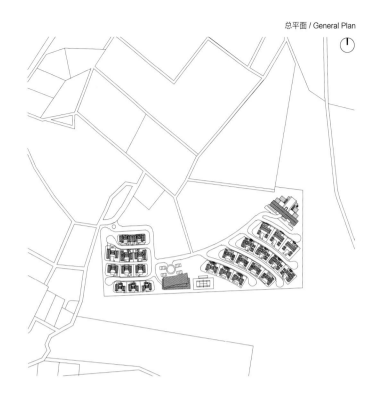

总平面 / General Plan

美国塞班岛马里亚纳别墅区

The Mariana Villa Area in Saipan Island USA
2011 ~ 2017 , Saipan , 17619 ㎡

本项目选址于塞班岛海军山区，地处西海岸中部半山间，风景秀丽、山林茂密，可眺望菲律宾海。设计充分尊重自然山体，并注重"海景优先"原则，利用山形的高差变化适度改造，做到让每栋别墅的主要房间都能观赏到海景，塑造独特的山地别墅；单体建筑的处理借鉴热带海洋性气候地区的建筑特点并融入对当地人文要素特征的思考；同时，以生活流线舒适和外部环境利用为依据，尽可能增大起居室、餐厅和主要卧室的观海面，庭院空间层次丰富，试图营造充满生机和社区交流的理想居所，使住户能够体验到全新的海岛居住模式，畅快而惬意地亲近自然。建筑造型采用新亚洲主义风格，利用不同形式的坡屋顶创造变化的建筑空间，以石材、白墙实木构架、造型窗格、照壁等精致的细部符号，塑造出典雅温馨舒适的别墅风貌。

This project is located in the Navy Hill district, Saipan. The site is at the mid-hill over the central part of the West Coast; it boasts beautiful landscape with dense forest, overlooking the Philippine Sea. With full respect for the natural hill the design adopts the principle of "seascape first"; taking advantage of the variation in the hill's altitude difference combined with moderate transformation enables the primary rooms in each villa to have access to seascape, building unique mountain villas. Full consideration is given to the architectural features of the region with tropical oceanic climate for the design of individual villa, incorporating the architect's thinking about the local human elements. Meanwhile, taking comfortable daily circulation and leveraging of external environment as the objective, the design increases the span with a view of the sea as much as possible in the living room, dining room and master bedroom of each villa, creating multi-layer space for the courtyard, in a bid to create an ideal residence that is full of vitality and convenient for communications with people in the community. As a result, villa residents can experience a brand new residential model in an island, where they can enjoy pleasant and comfortable moments with nature. The neo-Asiatic style is adopted for the building form: various types of pitched roof are introduced to create varied architectural space; delicate detail signs, such as stones, white wall, solid wood frame, figured panes, and screen wall are applied to deliver the elegant, warm and comfortable ethos of the villa.

别墅生成分析 / Villa Generation Analysis
外部实景 / Outside View

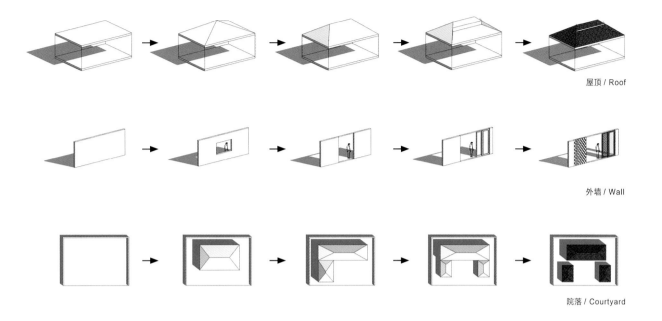

屋顶 / Roof

外墙 / Wall

院落 / Courtyard

人与自然通过科技在信息时代的整合
——上海浦东软件园设计

文 / 王振军

摘要： 本文通过介绍上海浦东软件园的设计实践，对软件工业园这种信息产业的新型建筑类型的基本设计要素，以及软件园的结构、空间环境、建筑形象进行了归纳总结，并对我国软件园的设计现状进行了分析。
关键词： 软件工业园；基本设计要素；自然；科技；人本主义

一、引言

软件工业园区与其他高科技园区相比共同具有人才密集，信息畅通，环境优美等特征。不同之处在于，软件开发最主要是依托于人的智慧和思维，而较少地依赖于高、精、尖的加工制造设备和测试仪器。因此，可以说软件园设计关注的不再是生产线的长度、精密设备适应的洁净等级及复杂的物流、货流等内容，而是更加关注人本身，关注人与自然环境的关系，关注人的体验。软件工业园应是"人本主义"建筑的最佳表现形式之一。

本文结合笔者在对国家三大软件基地之一的上海浦东软件工业园中的设计体验，对这一具有时代特征的设计类型展开论述。

二、软件工业园的特性分析

1. 软件工业园设计的三大基本要素：人、自然、科技无疑是软件园最主要的三大要素，软件园的功能、空间环境及形象设计均以此为出发点。软件园作为人、自然、科技的载体，通过对三者进行有效的整合，发现三者之间的秩序和结构以及它们之间相互适应的关系将是软件园设计所要探索的具体内容。在建筑发展过程中，一些基本要素随着时代的变化和建筑类型的变化，其地位和意义也相应发生了变化，技术要素在软件园设计中地位的变化正说明这一点。

2. 功能特性：软件业作为新兴的高科技产业日益成为各个国家新的经济增长点，政府和企业利用集中投资和优惠政策欲在短期内创造一个适宜软件企业成长发展的优越环境，特别是能够自我支持的创新环境。因此，软件园与游离分散的软件企业相比，应具有如下五大功能：（1）人才聚集功能；（2）群体协作功能；（3）内外辐射功能；（4）开发生产功能；（5）产品营销功能。

把以上功能分解归类可得出软件园区内建筑物的功能类型：（1）为软件企业提供技术支持的技术服务中心：包括数据通信中心、软件评测中心、电子出版中心、软件培训中心及产品演示中心；（2）生活服务中心：包括餐饮及娱乐、健身活动中心；（3）软件园的管理中心：行政、物业及招商引资管理及风险担保；（4）软件生产企业自身的工作空间：软件开发楼；（5）软件产品硬件化：系统集成楼。

3. 设计思想
（1）以人为本的总体指导思想：由于信息技术尤其是软件业所具有的"技术不可见性"的特点，技术对建筑的支配地位逐渐变弱，而软件的主体——人——无疑成为我们在软件园设计中关注的中心。

（2）可持续发展的生态设计观：一方面在注意总体规划整体性的同时考虑留出后续发展空间以适应软件企业的发展需求。另一方面要在建筑设计中充分考虑对自然环境的尊重和利用。

（3）开放的空间体系和朴素的形象观：强调建筑空间的流通和渗透以及与自然环境的交融，建筑造型追求朴素严谨的文化气质。

（4）因地制宜的设计手法：充分利用自然地貌，强调与自然的融合，塑造有层次的园区环境。

4. 空间及环境体系
（1）环境特征：位于美国旧金山东南的圣克拉拉山谷的硅谷，环境优美。而日本东京湾的软件开发区、英国的剑桥科学园区、中国台湾省新竹科技园等也都拥有高绿化率、低建筑密度以及丰富的自然地貌和田园风光。而这种先天性根本特征在具体的设计过程中又被建筑师进行了淋漓尽致的发挥和再创造。

（2）生态城市模式：拥有集中绿地为中心以及组团绿地和以簇状形态组合成的不同规模的建筑组群，园区空间注意生命感、人情化的表现。

（3）空间层次：注意人对空间的体验，不追求无谓的大尺度。注重从城市、园区、建筑的不同尺度来塑造空间。

三、设计实践：上海浦东软件园设计

浦东软件园是国家三大软件基地之一，位于上海浦东张江高新技术开发区 A8 地块。总用地面积 150000 平方米，总建筑面积 178000 平方米。

1. 总体方案构思

整个方案的立意和灵感是从分析地形及其与城市文脉的关系中产生，然后确定总体方案的指导思想：（1）注意与城市脉络在平面和空间上取得联系；（2）以生态城市观为指导创造出有利于激发人的智力性活动的自然环境。在空间上注重人的体验，创造宜人的有层次的空间；（3）可以持续发展的新型建筑设计观充分考虑园区的分期建设问题，保证后续开发时可自然衔接。

2. 设计方案介绍

（1）总体布置

中心绿庭加环形车道：方案根据用地狭长（南北平均长 750 米，东西为 250 米）这一根本性特点，园区中央沿南北向设有带形中心绿庭，内设步行系统，并结合水面、绿地、灌木、小山坡、雕塑，为园区创造一个休息、散步、交流的优雅环境。为保证中心绿带的宁静环境，本方案车行系统沿用地边界布置，生产用车、客车及消防均可在不进入中庭的前提下便捷通达各功能区。

（2）明确功能分区：本园区将建筑的功能归纳为三大组成部分即：

a. 管理、服务、信息中心；b. 软件开发楼；c. 系统集成楼。

由此我们在总体结构的节点部位布置管理服务及信息中心。以带形绿庭为中心，东侧紧邻城市主干道金科路将相对不怕干扰的系统集成楼沿街布置，而在相对较安静的居里路一侧布置软件开发楼。

（3）尊重天主教堂的存在，在布局上将教堂做为园区中心绿庭的端部对景来处理，从而增强了园区的人文内涵和对历史的尊重及对文化氛围的刻意营造，避免了高科技建筑通常所犯的孤芳自赏的通病。

3. 空间及环境设计

在软件园的一期入口处，会展中心、网络开发楼一起形成了第一个60米宽，围合而开放的广场，既满足了入园人流的集散需要，也是人们交往的一个重要活动场所，揭开了空间的序列。

穿过管理中心的过街楼，轴线上纵深的带形庭院是园区空间核心组成部分，它既是园区核心又起到划分不同功能区的作用。人们通过庭院进入各幢建筑，又从各幢建筑来到绿庭、休息、交流、相互启发。这是一个内向的围合性庭院，周围建筑错落有致，空间内外渗透，呈现积极高效的空间形态，建筑均以开放的形态面对内庭，软件开发楼一侧被赋予了一个活跃的曲线边界，隐喻了研究、开发这种高智力活动的特性，由此也增加了软件开发楼与绿化的接触长度。同时，考虑上海常年主导风向（东南风）及系统集成楼工艺生产对长度的要求，将建筑与南北成45°布置，有效地将风导入园区，同时从形体上也丰富了金科路的城市景观。

方案清晰地表达出这样的主题：要将整个园区建成一个与阳光、水、绿树相融的园区，一个生机勃勃充满自然状态的乐观主义氛围的园区。

4. 建筑设计

整个园区的建筑设计吸收了路易斯·康在美国宾夕法尼亚大学生物实验楼中的设计手法。这种手法将服务性空间做个性化处理，同时它与主楼部分拥有丰富的连接方式，从功能和形象均带来很大的优越性。面对入口广场做围合状处理，椭圆与阶梯状矩形呈均衡的体量左右布置，像一本被打开的书，同时表达接纳和吸收的含义。二侧旗杆、灯柱、绿化强化了进深方向的感觉，这些元素共同构成了软件园区的入口。主楼中123米长的直线型体量成了椭圆形会展中心和阶梯式矩形网络开发楼的背景，中段的底层架空把广场空间导入了中心绿庭（同时兼有消防通道功能）。

软件开发楼考虑到不同入园企业的需要，以不同的规模呈组团式布置，而建筑单体拥有共同的建筑符号系统，顶部空间是为软件开发人员准备的屋顶咖啡厅。平面的凸凹及门架处理均是为了在体量上造成宜人的尺度感。

系统集成楼将服务性房间在两端布置，中间留出大空间便于布置不同的工艺类型生产线。

用地中部偏南部位布置信息中心及服务中心，形体为四分之三圆环与矩形对接、中间的灰空间使绿化空间得以向南延续。

一期工程外装材料主要采用贴近自然的土黄色陶土毛面砖、乳白色涂料色带点缀，局部配以低反射率的蓝灰色玻璃幕墙，欲表达一种返朴归真的建筑意向。

5. 经济技术指标

（1）. 总体规划用地面积：15万平方米（一期用地面积为30000平方米）；（2）建筑占地面积：4.5万平方米（一期建筑占地面积8289平方米）；（3）总建筑面积：17.8万平方米（一期建筑面积33172平方米）。其中：管理、信息、服务中心共3.5万平方米；系统集成楼：4.2万平方米；软件开发楼：9.8万平方米。

四、设计体会

1. 软件园的创作不应该有固定模式，这种新型的建筑类型必将随着世界信息产业的发展变化及软件业本身的日益成熟以及当地的文脉、文化、国情及科技发展水平、投资状况等诸多因素的变化而具有不同的个性表现。软件工业园个性表现的追求应视为是丰富高新技术开发区的建筑艺术形象的契机，建筑师要珍惜这样的创作机会，切忌照搬。

2. 软件园这一特殊性质的高科技工业园区将会是一种充满生机的建设模式，它的功能系统、园区结构、空间秩序等，都需要我们在实践中进行不断的探索。但要注意的是，这一具有时代特征的建筑类型中所蕴含的人与自然、人与科技的关系这个最基本命题，将是我们在任何情况下的建筑创作中都应认真面对的永恒课题。

3. 从建筑艺术的角度来看，科技只是一种媒介，通过它而使人与自然更紧密地联系在一起。因为信息时代与工业时代相比人与技术的关系正从二元对立向以人为中心的方向转变。

参考文献
[1] 李大厦. 路易斯·康 [M]. 北京：中国建筑工业出版社，1993.
[2] [美] K·林奇. 项秉仁译. 城市的形象 [J]. 建筑师 19. 北京：中国建筑工业出版社，1984.
[3] 包锦明. 软件工业——高新技术产业建设的一种新型模式 [J]. 国际电子报，1996 (11/8).
[4] JA[J]. 日本：日本建筑师，1993 (3).
[5] 刘弘. 从机械空间到信息空间 [J]. 建筑学报，1996 (3).

Integrating Human and Nature via Science and Technology in the Information Era
——The Design of Shanghai Pudong Software Park

Text / Wang Zhenjun

Abstract: By introducing the design practice of Shanghai Pudong Software Park, the paper sums up the basic design elements for the software park — a new type of architecture in the information industry — and its structure, spatial environment and architectural image, and analyzes the status quo of the design of software parks in China.

Key words: software park, basic design elements, nature, technology, humanistic

I. Introduction

Compared with other high-tech parks, the software park has features such as density of population, smooth information flow and beautiful environment. The difference lies in the fact that software development depends primarily on human wisdom and thinking, but less on high-tech, precision, sophisticated manufacturing and testing equipment. Therefore, it can be said that the design of software parks focuses no longer on cleanness class required by the length of assembly lines and precision equipment and complicated logistics and cargo flow, but, more importantly, on human himself, on human-nature environment and on human experience. The software park should be one of the best expressive forms of humanistic architecture.

The paper, based on the author's design experience of Shanghai Pudong Software Park, one of the three largest software bases in China, examines the type of design with characteristics of the contemporary time.

II. Analysis of Characteristics of the Software Park

1. The three basic elements of the design of the software park: Human, nature and technology are undoubtedly the three main elements of the software park. The functionality, spatial environment and image design are all based on them. As the vehicle of human, nature and technology, the software park effectively combines the three. Finding the order, structure and mutually adaptable relationship between the three will be specific content pursued by designers of the software park. In the process of architectural development, some basic elements have experienced some changes in status and meaning corresponding to the changes of the times and architectural types. The change of the role of the technological element in the design of the software park has illustrated just that.

2. Functional characteristics: As an emerging high-tech industry, the software industry has increasingly become a new economic growth point in every country. The government and enterprises have employed concentrated investment and incentive policies with the purpose to create an advantageous environment favorable for the growth and development of software enterprises—especially a self-supportable environment of innovation —in a short time. Therefore, compared with separate software enterprises, the software park has the following five functions: (1) Talent gathering; (2) Group coordination; (3) Internal and external "radiation"; (4) Development and production; (5) Product marketing.

Breaking down and classifying these functions, the paper derives the different types of architecture in the software park by functionality: (1) Technical service centers on providing technical support to software enterprises, including the data communication center, software testing center, electronic publishing center, software training center and production demonstration center; (2) Life service centers, including the catering center, entertainment center and fitness center; (3) The management center of the software park: administration, property management, investment attraction and management, and risk guarantee;(4) Work space for software production enterprises —— software development building; (5) Hardware realization of software products —— system integration building.

3. Design philosophy:

(1) Human-centered guideline: because information technology, particularly the software industry, is characterized by "invisible technology", the domination of technology over architecture has gradually weakened. Human—— agent of software—— has undoubtedly become the focus of attention in our design of the software park.

(2) Ecological design concept of sustainable development: on one hand, while attention is paid to the wholeness of the overall design, consideration is also given to setting aside room for further development so as to meet the development demand of software enterprise; on the other hand, full consideration is given to respecting and utilizing natural environment in architectural design.

(3) Open spatial system and simple outlook on appearance: the design stresses the connectivity and saturation of architectural space and its harmony with natural environment. As for architectural form, a simple, well-organized cultural pathos is pursued.

(4) Design techniques adapted to circumstances. The natural terrain is fully made use of, with emphasis laid upon the integration with nature, to create a well-layered park environment.

4. Spatial and environmental systems:

(1) Environmental features: located in Santa Clara Valley in the southeast of San Francisco, the United States, Silicon Valley boasts a beautiful environment. The software development zone of Tokyo Bay, Cambridge Science Park and Hsinchu Science Park also have great greening rate, low architectural density, diverse natural landforms and picturesque scenery. The innate design features have been highlighted and recreated by architects in the actual process of design.

(2) Ecological city model: the model features a green space at the center, scattered green spaces and architectural complexes of different scales consisting of building clusters. When spaces of the park are designed, attention is paid to the expression of a sense of life and human feelings.

(3) Spatial arrangement: pay attention to human experience of space. Do not pursue large dimensions blindly. Create spaces in the different dimensional contexts of the city, the park and architecture.

III. Design Practice: the Design of Shanghai Pudong Software Park

Pudong Software Park, one of the three software bases in China, is located in Section A8 of Zhangjiang High- and New-Tech Development Zone in Pudong, Shanghai, with a total land area of 150,000 squaremeters and a total floor area of 178,000 squaremeters.

1. Concepts for the overall plan: The overall plan has derived conception and inspirations from an analysis of the relationship between the terrain and verve of the city. Then the guiding principles for the overall plan are determined: (1) Paying attention to establishing two- and three-dimensional connections with the city; (2) Creating a natural environment conducive to the stimulation of human intellectual activities, guided by the eco-city philosophy—— laying emphasis upon human experience spatially and creating well-layered spaces suitable for human; (3) Giving full consideration to the phased construction of the park with a new architectural design concept of sustainable development so as to ensure the natural connection with further development.

2. Introduction of the design

(1) Overall layout

Central green yard and circular driveway: The scheme, considering the fundamental feature of the narrow and long space (750 meters long on average from south to north and 250 meters wide on average from east to west), has laid out belt-like central green yard stretching at the center of the park from south to north. There is a pedestrian system, which combines water surfaces, other green spaces, bushes, hill slopes and sculptures to create a beautiful environment for rest, stroll and communication. To ensure the tranquil environment of the central green belt, driveways have been laid out near borders of sections so that production vehicles, passenger vehicles and fire trucks can access all functional zones conveniently without entering the central yard.

(2) Clear functional zones: the park has classified the architecture into three types by its functionality, namely a) management, service and information centers; b) software development buildings; and c) system integration buildings.

Therefore, we have laid out service and information centers at the junctions of the overall plan. With the belt-shaped green yard at the center, system integration buildings, which are not easily subjected to disruption, are placed along Jinke Road, the city's adjacent artery in the east. And the Software development buildings are placed along Juli Road, a rather quiet road.

(3) Accept the existence of the Catholic church. In layout, the church is treated as a sight opposite to an end of the central green yard, thereby enriching the culture of the park, showing the respect for history, creating a rich cultural atmosphere and avoiding the prevalent vice of superciliousness of high-tech buildings.

3. Spatial and environment design

At the entrance of the first-phase construction of the software park, the conference & exhibition center and network development building form an enclosed yet open square with 60 meters wide, which not only meets the need of in-going traffic flow, but also provides an important venue for interpersonal exchange and unfolds a spatial sequence.

Crossing the straddling building of the management center, you will get to the belt-shaped yard sitting on the axis, which is not only a core component of the park, but also serves as a border for different functional zones. People may cross the yard to enter different buildings and go to yard from the buildings to rest, communicate and inspire each other. It is a convergent enclosed yard surrounded by well-spaced buildings. With the inner and outer spaces interlinking, a positive, highly efficient form of space is demonstrated. All the buildings face the inner yard in an open form. On one side of the software development building, there is a lively curved border, suggesting the intelligence-demanding feature of research and development and adding to the length of contact between the software development building and the greening. Meanwhile, considering the commonest direction of wind (southeast wind) all year around in Shanghai and the demand for length of the production of the system integration building, the building faces the south-north direction at an angle of 45 degrees so that wind is effectively channeled into the park. Moreover, the cityscape along Jinke Road is also enriched.

The scheme clearly expresses such a theme: building the entire park into one where sunshine, water and green trees are harmoniously combined and a lively park saturated with a natural and upbeat atmosphere.

4. Architectural design

The architectural design of the park borrows techniques from the biological lab of Pennsylvania State University in the United States, which is designed by Louis Kahn. Service spaces are treated in a unique fashion; meanwhile they are joined to the main structure in diverse ways. This has given great advantages in function and image. The square facing the entrance is enclosed, with oval and stepped rectangles placed on the left and right in a balanced manner to resemble an open book, which signifies acceptance and absorption. The banner poles, lamp poles and plants on the two sides reinforce a sense of depth. Together these elements make up the entrance to the software park. The 123-meter-long straight structure of the main building becomes the backdrop of the oval conference and exhibition center and stepped rectangular network development building. The elevated middle section diverts visual attention to the central green yard (and serves as a fire exit).

To address the different needs of enterprises based in the park, the software development buildings have been arranged in clusters of different scales. All individual structures share a common system of architectural symbols. The space at the top is reserved for a roof-top cafe prepared for software development staff. The recesses and convex elements on surfaces and the treatment of door pockets are intended to create pleasant dimensions.

The system integration building has service rooms placed at the two ends so that spacious areas are reserved for assembly lines of different types.

In the south of the central area of the land, there is an information center and service center. With the former like 3/4 of a circular and the latter like a rectangle, the two join together. Together with the middle green space, they make the green space to extend southward.

The first-phase construction adopts mainly khaki-colored rough-surfaced tiles for external decoration dotted with white paint. At some parts, bluish green glass screen walls with low reflectivity are fitted. With the elements, the architects try to convey an intention of reverting to simplicity.

5. Economic and technological indexes

(1) Overall planned area: 150,000 squaremeters (with 30,000 squaremeters used in the first phase of construction); (2) Land area occupied by architecture: 45,000 squaremeters (with 8,289 squaremeters occupied by the first-phase construction); (3) Total floor area: 178,000 squaremeters(with the floor area of the first phase construction being 33,172 squaremeters). The management, information and service centers have a total floor area of 35,000 squaremeters, the system integration building a floor area of 42,000 squaremeters, and the software development buildings have a floor area of 98,000 squaremeters .

IV. Reflections on The Design

1. The design of a software park should not follow a fixed pattern. The new type of architecture will have different iterations with the development and changes of the world information industry, the increasing maturity of the software industry, and the changes of multiple factors such as local culture, national situation and level of scientific and technological development. The pursuit of self-expression in the software park should be seen as an opportunity to enrich the image of architectural art in the high-tech and new-tech development zone. Architects should treasure such opportunities and avoid blind copycatting.

2. The software park, a special type of high-tech industrial park, will be a mode of construction full of vigor and vitality. We need to carry out continuous exploration in its functional system, park structure and spatial order. What demands our attention is the fundamental subjects embodied in the architectural type with characteristics of the time, such as the relationship between human and nature, between human and technology. They are permanent subjects we should take seriously in architectural creation under all circumstances.

3. From the perspective of architectural art, technology is merely one medium, through which we can make human and nature more closely connected, because compared with the industrial era, the information era sees the human-technology relationship shifting from binary opposition toward the human-centered direction.

References

1. Li Dasha. *Louis Kahn* [M]. Beijing: China Architecture & Building Press, 1993.
2. Kevin Lynch. tr.Xiang Bingren. *The Image of the City* [J]. Architects(19). Beijing: China Architecture & Building Press, 1984.
3. Bao Jinming. *Software Industry —— A New Construction Model of High —— and New —— Tech Industry* [J]. Beijing: World Electronics, 1996 (11/8).
4. JA[J]Japan. JA. 1993 (3).
5. Liu Hong. *From Mechanical Space to Information Space* [J]. China: Architectural Journal, 1996 (3).

探索软件园设计的生态学途径
——上海国家软件出口基地规划设计

文 / 王振军 张会明

摘要： 生态系统是"一定空间内生物和非生物通过物质的循环，能量的流动和信息的交换而相互作用,相互依存所构成的生态学功能单元"。本文以上海国家软件出口基地为例，从生态学的角度探讨软件园的生态设计途径。

关键词： 软件园；生态设计；可生长性

随着世界信息时代的来临，信息产业的核心——软件产业已进入了一个高速发展时期。据统计，到目前为止仅限国家级软件产业基地总占地已达 5700 公顷，已建成和正在建设的建筑面积接近 2500 万平方米。在为我国软件园快速发展而欣喜的同时，我们不能不对这种用地量和建设量巨大的新型建筑类型从可持续发展的战略角度、从人与自然协调发展、从自然资源最大化利用和保护的角度投入更多的关注，而所有这些正是生态学所关注的问题。

生态是指人与自然的关系。"生态学认为，自然界的任何一部分区域都是一个有机的统一及生态系统(ecosystem)。生态系统是'一定空间内生物和非生物通过物质的循环，能量的流动和信息的交换而相互作用,相互依存所构成的生态学功能单元。'生态系统具有自动调节恢复状态的能力，达到能量和物质流动的平衡，及生态平衡"[1]。

本文以上海国家软件出口基地规划国际竞赛(第一名)设计为例，探索软件园的这种动态的平衡及软件园设计的生态学途径。

一、基地概况

上海国家软件出口基地位于浦东张江高科技园区的几何中心，西临园区主干道金科路，北侧为浮出地面的地铁并在西北角设站，隔祖冲之路为 IT 产业区，南临高科路，用地东南侧为医药基地和吕家浜水系。原规划城市道路接口共 4 个，已动工。

二、基地分析及问题的提出

整个园区是城市的一部分但又自成系统，所以其功能应该是在满足自身功能的前提下又是城市的一个有机体。随即产生了两个问题：

1. 对外：张江区绝大部分土地已被开发，各开发用地自成系统，城市区域配套设施仍严重缺乏。作为国家重点项目在促进企业发展的同时也带动整个区域经济的发展。(1) 满足园区规划要求的前提下，增加与城市相配套的设施；(2) 开发旅游资源促进区域经济的发展。

2. 对内：作为一个软件园本身应满足自身需要，人、自然、科技三大要素的整合。

人——软件园的运作主体——高智商的软件人才高地；自然——呈现生态、环保、高绿化率、低密度、低容积率的原生态自然环境——可使置身其中的人能将知觉和推理能力及其思维潜力更好地发挥；科技——现代计算机软硬件技术,构成软件出口基地技术服务和研发平台。规划结构、建筑型制能灵活地适应各种招商模式的需要，及建筑可生长性与地块的可生长性。

三、从以下几个方面来阐述解决途径

1. 功能分区的生态学途径

(1) 用地方位的考虑

a. 沿金科路与祖冲之路交叉口综合地铁站出入口处布置外向型的商业服务配套设施，充当一部分城市功能。

b. 将居住部分置于靠近祖冲之路和多媒体公园一侧，其作息时间正好与地铁运行时间错开，互不干扰，靠山面水，景色宜人。园区酒店置于用地中央为园区的视觉中心，既服务于园区又为充当城市功能。

c. 将软件研发置于沿金科路和高科路，既便于入驻公司展示形象和方便出入，又便于自成一体形成高科技园区形象。

d. 将培训中心置于吕家浜畔，景色宜人，环境静宜，适于教学培训。

e. 生活服务配套置于中心水景四周，表现出强烈的亲水性，同时符合 C·亚历山大提出的人对邻里的认知范围 [2]，对研发和居住均有较适宜的服务半径。

注：C·亚历山大认为邻里认知范围直径为 274 米，步行可达行角度考虑在 200 米以内和 5 分钟步行距离。

(2) 最佳的自然采光

科研大部分高度设计在 12 米～24 米之间增加亲地性，科研区建筑间的间距在 50 米左右，SOHO 区南低北高，其布置方式有助于建筑拥有最佳的自然采光，因而减少能源的耗费。

(3) 最佳的自然流畅的空气流和最佳的景观视野。

点、线、面结合的总体布局有利于空气的流动，调节区域气候起到一定的积极作用。中央水景公园周边建筑的布置确保每一栋建筑均至少有一面景观面向中心。

(4) 废物管理系统及物料供应系统。

基地在每个地块地下车库内设立整体考虑的废物处理管理系统，办公及生活垃圾实行分类袋装化处理，集中于垃圾间，由专人管理，每日由环卫部门清运。

物料供应即研发所用磁盘、光盘、纸张及生活配套服务所需物流也从相应地下车库输入到各部分的地上部位。

2. 景观设计的生态学途径

景观与土地的利用相关，使某一区域区别于其他区域的构成特征是景观(landscape)，它是多元素组合，包括田野、建筑、山体、森林、荒漠、水体以及住区。景观并不只是像画一般的风景，它是人视线所见各部分的总和，是形成场所的时间与文化的叠加与融合，是自然和文化不断雕琢的作品。

(1) 景观模型

为了获得对自然过程、人们计划或活动的认识，对用地尺度进行了分析和研究提出了两种景观模型，提出的模型需能保证目标的完成。

模型 1：风和日丽的自然之场(脉络)

在人和科技之风的催化下舞动起来，自然之力渗透到了基地各个角落，一颗充满智慧和灵性的生命之水(湖)由此诞生。

模型 2：景观风水模型

"生生之气"即阴阳冲和之气，落实到形则为藏风聚气之地，具体表现为山势连绵起伏，水系回环有情、源远流长，土厚植茂；空间上讲求层次感和多等级的分形系统。

空间构想基于这种构思，所有地块呈向心布置，空间趋向呈四周高向中心逐渐降低，从金科路和祖冲之路交叉口渗入的城市空间和从吕家浜渗入的水景空间作为步行系统和景观轴线汇聚在基地中心。东北区域的大型集中绿地微微隆起，渗入湖面，湖面上智慧岛中的酒店公寓主体 45 米，局部 50 米，成为园区的制高点和园区视觉中心，顶部设有园区的数码俱乐部。整个园区呈现从城市空间到自然中心的"层层渗透"和从自然中心向城市空间辐射的积极状态。

(2) 景观轴线

轴线的设计使得园区与城市有了更好的对话。在轴线上行经一个逐渐放松的过程体验，既是功能性的又是视觉上的。

(3) 景观的特性

评价方案的景观特性可总结为：a. 整合性：向心性和渗透概念保证了景观的整体特性；b. 外围外向性的绿带和水景与内环内向性的对比；c. 向心性的楔形空间和视觉走廊保证了基地景观具有层次感的同时又有一定的景观联系性。

(4) 水景系统

水景系统由中心湿地和环形水系构成，两个水系交汇于吕家浜，环形水系平行于环路呈线性状态(4.5 米宽)，在研发地块可渗透入组团内形成不同特色的组团景观，对组团小气候进行了有效的调节，线性水系蜿蜒到地铁地下商业中心玻璃采光顶上，闪闪发光，充满灵性之美。

中心湿地在入吕家浜处设有一个"数码岛"，使从西向东而流过的城市水系吕家浜一进一出与中心湿地交融在一起，湿地由此变为活水。亲水建筑平台、挑台、码头等形成凸凹变化湖岸。舒缓与急促、自由与几何配以竹林、垂柳、玉兰，勾画出惬意和抒情的江南景色。

(5) 水景维护和保持

由于张江镇地下水位较高，基地场地浅部地下水属潜水类型，主要补给来源为大气降水。其地下水位埋深通常为 0.50 米左右。本基地浅部地下水对混凝土无腐蚀性。鉴于此，中心水面和线性水景的建造费用即可大大降低，通过调整水底标高，局部采用水泵控制水的流向，即可满足水位常年保持不变的目的，并实现自然换水。基地中心设有中心湖面，水深约 2 米至 2.5 米，可设小型湖泊游览设施，在湖面上小憩，将软件进出口基地尽收眼底，与基地环形路相并行，有环形水泵与中心湖面相连。中心湖泊水来源于吕家浜河水、湖泊周边雨水及地下水，湖泊可充分利用吕家浜自然水面与中心湖泊水面重合，利用水体生态系统、水底栖物植物的生长，靠静态标高差使水体流动，参与循环，尽可能避免"死水"，保证水体自然净化。环形水系为浅滩水可采用河流推进式，利用其推力达到循环。基地内植物及绿化带浇灌可取自中心湖水。平时还可通过人工喷淋(如景观喷泉)来调节水中气体成分，维持生物生长需要。

池底开挖后经夯实除湖边侧壁需做建筑处理外无需做特别处理，水质保持采用了湿地概念优化了水处理方式。

3. 交通设计的生态学途径

(1) 机动车道

充分注重现有城市规划给予基地四周的道路接口位置，在进入基地后连接成一个 24 米宽环形主干道，该环形主干道将占园区建筑面积 68% 的建筑组团直接连接起来，在主干道与沿湖生活配套服务建筑之间设有 8 米宽的次干道，主要解决生活区的物流和消防问题。在每个研发地块，均可以保证环形单向车道环绕。

(2) 步行系统

在金科路与祖冲之路交叉口(地铁站)通向中心湖面方向与景观轴一起为步道系统，同时在沿湖周边设有滨水步道系统。

(3) 停车系统

基于张江地下水位高、表层土质软等地质情况，建筑需打桩到 5 米以下，同时为设地下室具备了充分条件，本案中所有停车设于地下，既保证了地面的绿化率并有效地利用了地下空间。

(4) 消防系统

外环研发地块有高层建筑分布，内环酒店式公寓主体高度 45 米为高层建筑，地块内均设有环形机动车道(局部为隐蔽式消防道)或沿两个长边设置消防车道。其他建筑均为多层建筑，建筑长度均控制在 150 米内，超过时均在其中部设有净高净宽大于 4 米的消防车道。

4. 开发模式的生态学途径

(1) 地块和组团分析

规划架构确定后自然形成内环和外环两大部分，其用地的连续性适应了入驻企业的不确定性。生长式建筑为开发商提供了开发的灵活性。

(2) 土地开发强度概述

因研发是在金科路沿线，区位价值明显，为各入驻公司首选用，研发区在满足最低绿化率的前提下应提高用地的开发潜力，增加开发强度，居住部分布置在多媒体公园和中心水面之间能较好保证自然间距和绿化率，商业部分与园区开放空间结合，与树林和广场结合，开发强度宜考虑适中为宜。

地铁站处的地块，商业价值较高，应充分发挥其潜能。中央湖面的浮游之岛在现阶段做集中绿地是环境极佳的休憩之地，随着形势的发展和用地的需求增加，可作为一个极品地块用作研发用地。

(3) 分期开发

分期开发应考虑衔接顺畅，分期完成后整体统一、自成一体，同时交通、景观功能配置又很完善。

(4) 土方平衡

现有基地地形平坦，几乎无标高变化，多媒体公园隆起的山坡用土取自中心湖面挖方和研发和居住部分地下室的挖方，土方可基本取得平衡。

(5) 植被

规划区内植被茂盛，主要分布在水系丰富的区域，树种有玉兰、杜鹃、毛竹等。规划应尽可能将大部分茂密、丰盛、长势良好的树木予以保留，使其适合园区生态功能要求，保护及维护本规划区的生态体系。

四、总结

软件园项目中强调人与自然的和谐，强调园区景观环境对人精神愉悦的诉求，强调园区生态性、可持续发展性，这些已越来越清晰的成为软件园价值体系的基石。生态学途径，这种面向未来的设计方法无论对建筑师还是规划师都提出了新的设计要求和挑战。我们深信软件园这种信息时代特有的产物会伴随着我国软件园的发展、伴随着生态学方面研究的深入最终会成为信息时代一道靓丽的风景线。

参考文献

[1] 李小凌, 俞孔坚. 生命的景观 [M]. 北京: 中国建筑工业出版社, 2004 (4).
[2] 俞孔坚, 李韶华. 景观设计: 专业学科与教育 [M]. 北京: 中国建筑工业出版社, 2003 (9).

An Exploration into Ecological Approach for Software Park Design
——Planning and Design of Shanghai National Software Export Base

Text / Wang Zhenjun, Zhang Huiming

Abstract: Ecosystem is "an ecological functional unit of biota and non-living things within a certain space that interact with and inter-depend on each other through matter circulation, energy flow and information exchange". By taking Shanghai National Software Export Base (the base) as an example, this paper studies the ecological design approaches for software park from the perspective of ecology.

Keywords: software park, ecological design, growth

As the advent of the global information era, the software industry, which is the core of the information industry, has entered a stage of high-speed development. According to statistics, by so far, China's state-level software industry bases cover an area of 5,700 square hectometers, and the floor area of the completed and under-construction buildings of software parks has approached 25 million squaremeters. It is no doubt that we should rejoice at the rapid development of software parks in China, nevertheless, we have to pay more attention to the new-type buildings that demand great land and construction workload from the strategic perspective of sustainable development as well as the perspectives of harmonious development between man and nature, and maximum utilization and protection of natural resources, which fall in the scope of ecology.

Ecology means the relationship between man and nature. Ecologists believe that "any part of the nature is an organic, unified and ecological system (ecosystem). Ecosystem is 'an ecological functional unit of biota and non-living things within a certain space that interact with and inter-depend on each other through matter circulation, energy flow and information exchange.' Ecosystem can be automatically regulated and restored, so as to achieve the balance between energy and material flow, and ecological balance".[1]

This paper takes the planning and design of Shanghai National Software Export Base (the first place in the international competition) as an example to explore the dynamic balance of software park and ecological approach for software park design.

I. Overview of The Base

Located at the geometric center of Zhangjiang Hi-Tech Park, Shanghai National Software Export Base borders Jinke Road (the main trunk road of the Park) in the west and a subway station in the northwest, across the Zuchongzhi Road from the IT industry zone. It borders Gaoke Road in the south, with a biomedical base and the Lvjiabang river system in the southeast. The construction of the four road intersections that were previously planned have started.

II. Analysis on The Base and Relevant Problems

As well as a part of the city, the entire park has created a system of its own. Therefore, it shall be an integral part of the city under the precondition of meeting its functional requirements.

1. Most lands of Zhangjiang Development Zone have been developed, and the development areas are separated from each other and plagued with severe lack of urban infrastructures. It is a key project of China, and its development shall drive the development of the local economy. (1) Increase supporting facilities under the precondition of satisfying the requirements of parking planning. (2) Develop tourism resources and promote local economic development.

2. The software park itself shall meet its own needs and achieve integration of human, nature and technology.

People: operator of the software park and software talents with high intelligence quotient.

Nature: original ecological environment featuring environmental friendliness, high green coverage, low density and low plot ratio, allowing people to make better use of their perception ability, inferential capability and thinking potential.

Technology: modern computer software & hardware technologies that set up a platform of technical services, research and development of the software export base.

The planning structure and architectural style can meet the needs of investors and are conducive for growth of building and land parcel.

III. The Solution is Expounded from The Following Aspects

1. Ecological approaches for functional zoning

(1) Location

a. Arrange outward commercial service facilities at the intersection between Jinke Road and Zuchongzhi Road and the entrance and exit of the subway station, and make the facilities act as a part of urban functions.

b. Arrange the living area on the side of Zuchongzhi Road and the multimedia park, where has an enchanting scenery, so as to ensure that it will not be disturbed by subway operation. Located at the center of the land, the hotel is the visual center of the Park, and both serve the Park and act as a function.

c. Arrange the R&D Area along Jinke Road and Gaoke Road, making it convenient for easy access and image show of relevant companies and the Park.

d. Build the training center on the bank of the Lvjiabang River, which is suitable for teaching and training for its beautiful scenery and quietness.

e. Arrange the life service facilities around the central lake with beautiful waterfront sceneries, in conformity with the neighborhood cognition theory that C. Alexander put forward [2] and near the R&D and living areas.

Note: C. Alexander believes that the diameter of neighborhood cognition is 274 meters, and the walkable distance shall be less than 200 meters (or five minutes' walking distance).

(2) Optimal natural lighting

The heights of most scientific research buildings range from 12 meters to 24 meters; the spacing between buildings at the R&D area shall be about 50 meters; the soho area declines southward, in this way allowing the buildings to have the best natural lighting and reduce energy consumption.

(3) Optimal natural and unobstructed air flow and optimal scenic view.

The general layout of point-line-surface integration is beneficial to air flow and plays a positive role in regulating regional climate. Arrange the buildings around the central waterscape park and ensure that at least one side of such buildings faces the park.

(4) Waste management system and material supply system

Set up an integrated waste treatment management system in the underground garage of each land parcel of the base, treat the office and household refuses in a classified way, bag the refuses and deliver them to the refuse room in a centralized way, designate a special person to manage the refuses, and environmental sanitation department clear the refuses on a daily basis.

Transfer such materials as magnetic disk, CD and paper for research and development and materials for life services from corresponding underground garages to the above ground.

2. Ecological approach for landscape design

Landscape is related to land utilization and differentiates a certain area from other areas. It is a multi-element combination, including field, building, mountain, forest, desert, water body and residential area. Landscape is not just picturesque scenery, but is a combination of all visible features of an area of land. It involves the cultural overlay and integration of a certain site, and natural and cultural works that are constantly "shaped".

(1) Landscape model

In order to gain an understanding of natural process, planning and activities, this paper analyzes and studies the land use scale and puts forward two types of landscape models, which may help attain the above goal.

Model 1: Natural Field

As a result of the impacts of human and technology, it exerts the force of nature to every corner of the base, giving birth to a lake that is full of wisdom and intelligence.

Model 2: Landscape "Feng Shui" Model

The model means the harmony between "Yin" and "Yang". Physically, the Park is a place endowed with excellent conditions, namely the rolling hills, winding rivers, fertile land and lush vegetation. Spatially, it is a hierarchical system with a sense of depth.

Spatially, all land parcels of the base decline gradually towards the center. The urban space that penetrates from the intersection between Jinke Road and Zuchongzhi Road and the waterscape space that penetrates from Lvjiabang River converge at the center of the base as pedestrian system and landscape axis, respectively. The large green space in the northeast area of the Park rises slightly and extends to the lake. The main building of the hotel apartment on the Intelligent Island in the lake is 45 meters high, and 50 meters high in some parts. As the commanding height and visual center of the Park, the main building has a digital club on its top. The entire Park penetrates gradually from the urban space to the natural center and radiates from the natural center to the urban space.

(2) Landscape axis

The design of axis makes better dialogues between the Park and the city. It is a relaxing experience to walk on the axis, both functionally and visually.

(3) Landscape characteristics

The landscape characteristics of the solution can be summarized as follows: a) Conformity: centrality and percolation concepts ensure overall characteristics of the landscape. b) Contrast between the outer greenbelt and waterscape on the periphery and the inner features of the inner ring. c) The wedge space with centrality and visual corridor ensure that the landscapes of the base have a sense of depth and a certain connectivity between each other.

(4) Waterscape system

The waterscape system is composed of the central wetland and the annular water system, which converge at the Lvjiabang River. The line-like annular water system is parallel with the ring road (4.5 meters wide), penetrates into the cluster at the R&D land parcel, creates landscapes with different features and can effectively regulate the microclimate of the cluster. The water winds its way through the glass day-lighting roof of the underground business center of the subway station, dazzling with brilliance. The central wetland has a "digital island" at its intersection with Lvjiabang River. The island makes Lvjiabang River that flows from west to east converge with the central wetland, thus turning the wetland into a lake of living water. The waterfront building platforms, cantilever platforms and wharfs form a graceful lakeshore with gentle lines, sharp curves as well as free and regular visual elements. Dotted with bamboo, weeping willow and magnolia, the base has pleasing and picturesque scenery.

(5) Waterscape maintenance and retention

Zhangjiang Town has a relatively high underground water level. The shallow underground water of the base is of submersible type and is mainly replenished by atmospheric precipitation. The buried depth of groundwater is usually about 0.5 meter. At the base, the shallow ground water is corrosive to concrete. For this reason, the expense for creation of the waterscape on and around the lake can be significantly lowered. By adjusting the bottom level and local pump control of water flow, the water level can be kept unchanged and natural water exchange can be achieved. The center of the base has a central lake with a depth ranging from 2 meters to 2.5 meters, where small sightseeing facilities can be provided for people to relax on the lake. Parallel with the ring road of the base, the ring water pump is connected with the central lake. The water of the central lake comes from Lvjiabang River, rainwater and underground water around the lake. Lvjiabang River is connected with the lake, therefore, it is possible to make the water circulate by means of the water ecosystem, growth of benthos and static elevation difference, so as to avoid "dead water" and ensure natural purification of the water body. The annular water system shall be shallow and its circulation shall be driven by the river water. The water of the central lake may be used to irrigate the plants and greenbelts in the base. At ordinary times, artificial sprinkling (such as landscape fountain) can be used to adjust the gas composition in the water and ensure biological growth.

Except for the banks, other parts of the lake need no special treatment after the lake bottom has been excavated and tamped, and the wetland concept is used to optimize the mode of water treatment for water quality maintenance.

3. Ecological approach for traffic design

(1) Road for motor vehicles

Pay full attention to the locations of the roads around the base in the existing urban planning. The roads form a ring trunk road with a width of 24 meters after entering the base, directly connecting building clusters with a floor area accounting for 68% of that of the total building clusters; a secondary trunk road with a width of 8 meters is built between the trunk road and the life service buildings around the lake, for the purpose of meeting the logistics and fire control needs in the living area. Each land parcel for research and development is surrounded by a one-way annular lane.

(2) Pedestrian system

Build pedestrian systems at the intersection between Jinke Road and Zuchongzhi Road (subway station) facing the central lake and the landscape axis, and set up a waterfront pedestrian system around the lake.

(3) Parking system

Drive the piles into the earth at a depth of at least 5 meters in consideration of the geological conditions of high underground water level and loose surface soil in the area, and create sufficient conditions for building basements. In this solution, all parking lots are arranged under the ground, not only utilizing the underground space effectively but also making room for increasing the landscaping ratio above the ground.

(4) Fire system

There are high-rise buildings on the R&D land parcel on the outer ring; the main building of the hotel apartment on the inner ring is a high-rise building with a height of 45 meters; all land parcels have ring roads for motor vehicles (concealed fire lane at some places) or fire lanes along the two sides of the base. All other buildings are multi-storey buildings with a maximum length of 150 meters. If exceeded, a fire lane with both of its clear height and width greater than 4 meters shall be set up.

4. Ecological approach for development mode

(1) Analysis on land parcel and cluster

It forms two major parts (inner ring and outer ring) naturally after the planning framework has been determined, and the continuity of land use suits the uncertainty of companies. The growth-type buildings may provide the developers with development flexibility.

(2) Overview of land development intensity

Located along the Jinke Road, the R&D Area is the first choice for all companies due to its clear geographical advantages. Under the precondition of achieving the lowest landscaping ratio, the R&D Area shall fully tap the development potential of the land and increase the development intensity, and the living area shall be arranged between the multimedia park and the central lake, so as to ensure natural spacing and landscaping ratio, integration between the business area and the open space of the Park, and integration between woods and the square. Ideally, the development intensity shall be moderate.

The land parcel of the subway station has high commercial values, which shall be fully utilized. In the present stage, the floating island at the center of the lake is designed as a gorgeous green field for relaxation, which can be turned into an ideal area for research and development as the situation develops and the land demands increasingly.

(3) Phased development

The phased developments shall be linked smoothly and form a whole upon completion, with perfect traffic and landscape functions.

(4) Earthwork balance

Currently, the base has a flat terrain and has almost no elevation change, the earthwork of the hill in the multimedia park comes from excavations of the central lake and the basements in the R&D area and living area, and the earthwork can be basically balanced.

(5) Vegetation

The planned zone has lush vegetation, which are mainly distributed in areas with plentiful water systems, including magnolia, rhododendron and moso bamboo. Most thick, lush and healthy trees shall be retained as far as possible, so as to meet the ecological functional requirements of the Park and protect and maintain the ecological system of the planned zone.

IV. Summary

The software park project emphasizes the harmony between human and nature, spiritual pleasure in the landscape environment as well as ecological features and sustainable development of the Park, which have become definitely the cornerstone of the value system of the software park. The ecological approach, as a future-oriented design method, has put forward new design requirements and challenges for architects and planners. We are convinced that, software park, as a specific product in the information era, will ultimately shine brilliantly as it develops in China and the ecological studies deepen.

References

1. Li Xiaoling, Yu Kongjian. *The Living Landscape* [M]. Beijing: China Architecture & Building Press, 2004 (4).
2. Yu Kongjian, Li Shaohua. *Landscape Design: Professional Discipline and Education* [M]. Beijing: China Architecture & Building Press, September 2003 (9).

回 归 本 质
2011年中国建筑学会学术论坛上发表的主旨演讲

文 / 王振军

沙特馆在2010上海世博会上有两个非常突出的特点：第一它是唯一一个由中国团队独立完成建筑设计全过程的一个场馆，包括室内和景观。而这个不重要，重要的是第二点，即这个馆从开幕到现在，是一个逐渐由冷变热的典型过程，这个过程很耐人寻味。开幕之前没有什么媒体报道，我们的团队也是诚惶诚恐，工作两年半，怕大家不喜欢它，说实在当时心里还是捏了一把汗，而开幕以后一天比一天火，高峰排队时间一度达到九小时以上。

在这个由冷变热的过程中，也正好是我们作为一个建筑师，特别是全过程参与其中的建筑师进行反思和学习的过程，也是思索建筑学当中一些比较本质问题的一个过程，比如说形式和内容的关系，形式和功能的关系，展览建筑或者说建筑的本质是什么，检验建筑好坏的标准应该是什么等等。

经历这样一个世界性的重大建筑事件，对于我而言是非常难得的学习过程，也无疑会对我今后的职业生涯产生重大的影响。下面我就把自己在这个过程中的感受和收获给大家做一个汇报，也试图去揭示一些这场盛大的建筑选美表象背后的本质，而我认为这也是举办这场规模宏大的人类盛会对建筑师来说最大的价值所在。

一、世博会国家馆的功能本质

纵观159年的世博史，在20世纪60年代西方经历了经济危机后，世博会在举办的时候已经从技术的博览变成文化主题的展示，或者说是国家形象宣传的平台。从这一届世博会，我们也能够看到技术的展示或技术的实例已经不具备唯一性，也就意味着失去了展示性。例如，不管是发达的还是不发达的国家都用了触摸电视，你那边有多媒体，我这边也有。因而大家把更多的精力放到了文化的展示与交流上，所以说世博会外国馆主要的目的是为了展示每个国家多样性文化。在做竞赛之前，我们通过对世博会历史的了解，也深刻地认识到这一点。事实上，做外国馆其实是在做一个展示该国的文化容器，把所在国的文化最大限度地或者说最大容量地放进去，这是最重要的。

通过设计沙特阿拉伯馆，让我对沙特阿拉伯这个民族有了更深的了解。我觉得这个民族是一个很睿智的民族，为什么这么说？因为他是世博会42个外国馆中唯一一个将其拿出来进行全球方案招标的国家，其他所有展馆的招标都是只限定于本国建筑师。从这一点来说，我十分佩服他们，佩服他们的勇气、自信和眼界。沙特城乡事务部的副部长也认为这样的决策是正确的。而这种魄力也确实取得了一个很好的结果和回报，世博期间许多媒体评价说沙特是外国馆里边最大的赢家。

二、世博国家馆的建筑本质

在2007年11月的第一轮竞赛中，我们提交了三个方案，分别是月亮船、阿拉伯魔方和快乐城堡。第一轮的时候月亮船、阿拉伯魔方均入围，最后月亮船方案中选。在整个方案设计过程中，我们是基于对沙特阿拉伯馆功能的分析和判断来展开的，既然作为一个文化的容器，对一个国家文化、政治、教育、城市等都要有一个深入的了解。在第一轮不到两个星期的投标时间中，我们大概花了一半的时间来研究这些，为此我们还特别聘请了北大的付志明老师作为顾问，现在想来用这一周的时间是非常值得的，为我们更进一步构思方案、判断设计走向起到了至关重要的作用。

月亮船这个构思来源于一千零一夜中关于月亮船的故事，想象其从遥远的中东沿着海上丝绸之路飘浮到上海湾，同时也寓意了中沙深厚的友谊。在构思确定后，怎么把文化的内涵进一步拓展而使其变得更加丰满就显得极为关键。

首先我们来看一下建筑的整体布局，地形是长方形，但是船头扭转角度，实际上是朝向麦加方向，麦加在阿拉伯民族心目中是一个圣地，这种轴线的扭转是表达对阿拉伯文化的尊重，预示着展馆强烈的宗教文化精神，这也得到了业主充分的赞赏。而景观上的设计也同样是在充分研究阿拉伯的文化特质后，从中提取相关元素并植入到建筑设计当中的。

在建筑的构思确定以后，就应围绕其本质功能进行深化。阿拉伯馆就是一个展馆，就是一个博物馆。博物馆设计的本质应该是空间体验的塑造。

在对世博会现场情况的研究中，我们发现排队等候是历届世博会的常态。由于规模所限，排队是不可避免的（但的确没想到实际能达到八、九个小时），当时查资料最长时间是两个半小时，我们就按照最长时间来设计等候的场地。考虑在等候的过程中怎么营造气氛，使其变成正式参观前的预热，比如在等候空间围绕舞台，排队时可欣赏阿拉伯歌舞等表演。另外，当地阳光和温度曲线图对我们方案的影响是比较大的，从上海常年太阳辐射的曲线图可以看出，世博会开幕的5到10月份是上海气候最不舒适的时间，既热又晒又潮。因此，我们特意把月亮船架空，从而营造出一个阴凉的等候空间。

第二个是主题参观，在漫长等待之后观众进入到第二个舞台，这是整个馆最大的创新点。当时最早的第一轮方案中楼板是铺满的，到第二轮的时候，感觉这种传统的参观方式连我们自己都觉得没有新意，的确传统的建筑已经没法满足在高科技时代的人们的关注点，那个时候《阿凡达》这部电影也对我们有一定的启发。生活在高科技时代中的人们需要一种更逼真、更融入或者说是更震撼的一种体验，传统"看"与"被看"的二元并置模式，不管什么形式，是放一个平板电视、一个雕塑、还是放一个机器人，总归是传统的参观方式。基于这一点考虑，我们干脆索性就把楼板抽掉，让参观者悬在空中，将船的内壳当屏幕。有一个媒体描述说"在这里没有固定的座位，你可以大步走在传送带上,跨越千山万水,尽情去触摸沙特的文明",淋漓尽致地说出了我们当时的设计的意图。其实简单地说,就是让观众换个方式看电影。

第三个是总览,也就是极具阿拉伯风情的屋顶花园,在这里观众可以俯瞰整个世博园,对5.28平方公里的整个园区有一个宏观的把握,很多观众在这里拍照留影。之后,通过螺旋式的坡道从中庭盘旋下来,在这种从室内外空间的相互转换中,观众对整个参观过程又有了比较深刻的回味。

三、绿色建筑的本质

在设计沙特馆时我们并没有刻意地去做这个概念。应该说节能是设计的契机,也是一种约束力,而这种约束力要心甘情愿的接受它,把它融入设计当中。所以我们在做建筑造型空间的时候,就争取将建筑美学、建筑的表现力与绿色建筑的概念融为一体。节能就是最大化地利用自然资源和相应技术以减少能耗。那种节能上的高技术则意味高成本,这也是一个不争的事实。

在沙特馆设计中我们的具体措施是:

1. 上大下小的造型使展馆几乎回避了日晒的影响,船体外壳因应功能的需要不设外窗,有效地减少了建筑与外界的热量交换。
2. 针对上海的亚热带季风气候,且5到10月世博期间又正值峰值段,我们选择将展馆架空来营造凉爽舒适的室外等候和表演空间,使参观的形式和内容前移。
3. 展馆内的交通核心作倒锥形处理,使内外环形坡道得以充分利用自然光,从而减少对人工照明的过度依赖。
4. 充分利用自然通风和上下温差,用天井垂直拔风原理,促进和改善了馆内的空气流通。
5. 镂空的外墙在结合阿拉伯风格营造的同时起到了遮阳的作用。屋顶的阿拉伯花园在展示沙特风貌的同时又有效地改善了展厅屋顶的隔热性能。

四、体会

第一,建筑的本质

我们的先哲老子在2000多年前讲到"有无相生"理论的时候,已经认识到做建筑其实是在做空间。在建一个房子的时候,空的地方是最有用的,我在做沙特馆的时候,深切体会到前人的伟大,我们在做建筑就是在做空间,空间就是主角。而在此应强调的是人又是空间的主人,一个空间再宏大,也是供人来使用和体验的。这就是挪威哲学家N·舒尔茨所提说的场所精神,场所是要人来体验,人和空间组合在一起才能产生真正的建筑体验。这应该是我最大的感受,现在有人说,当代建筑已进入一个自由时代或者说试验的时代,无论是什么样的观点,空间在建筑的主角地位是毋庸置疑的,人又是空间的主角,这样一个命题同样也应该是永恒不变的。

第二,建筑感知度

感知度这一名词是信息学的一个名词,指人对信息的接近程度,是对建筑师传播出的信息的感知程度。沙特之所以得到观众认可,是因为它的感知度比较高。从《一千零一夜》的故事中提取出的元素,还有丝绸之路等大家非常熟悉的,在世博期间开会的时候跟国外建筑师聊起来也是广为熟知的,大家小时候都看过《一千零一夜》。所以,一个建筑的构思应该具有比较高的感知度,在建筑层面即是寻找一个高感知度的元素来引起共鸣,又不能太晦涩,尤其是世博建筑等公共建筑的创作概念更应如此。

第三,对建筑创新的理解

设计是对社会变革的一种反映,这句话说得非常到位。沙特馆的室内流线设计是观众比较喜欢的,可能是因为我们这种展示的方式呼应了高科技时代对参观博物馆的体验需求。它改变了传统二元并置这种比较固定的模式,创造了一种更逼真、更融入或者说是更震撼的参观体验方式。把楼板抽掉,将流线架空处理,架在屏幕之上,这种模式的确颠覆了以往的体验。从这点上来说,建筑师创作的精力应放在对空间体验的创新上,而不是将太多的精力耗费在时尚的外形上,时尚的东西容易过时,本质的东西才是永恒的。

第四,建筑的社会属性

建筑师是不应把建筑当成塑造其个人特征的手段。何镜堂院士在演讲中说,中国所有人都是中国馆的业主,这句话是在说建筑,特别是针对世博会建筑的社会属性。由于其社会属性就不能用其他艺术形式来雕塑,比如纯用美学的方法来评价建筑;而且它还是一种空间的艺术,也有很多文化的使命。所以,从这点来说,建筑已经脱离了纯建筑师个人的美学范畴,而对建筑的价值判断也已经融入了业主,包括大众等各个层面人士的意见和判断。

第五,好的建筑应该是真实的建筑

我经常问自己,什么是好的建筑?我总结了几条。第一是在设计当中投入了真实的感情;第二是概念清晰,大家能引起共鸣;第三是形式和内容高度统一;第四是空间体验生动感人;第五是要关照文化传统,因为建筑承载着传承文化的作用,要创作体现文化内涵的建筑,这是建筑应担负的使命。

第六,世博建筑功能的拓展

在往届的世博国家馆中,贵宾接待仅考虑少量的接待情况,接待规模最大的也不超过100平方米,且未在其设计上投入更多关注。但是这次沙特方特意提出1000平方米的需求,当时我也不太理解为什么要做这么大,这样会把主馆的形态削弱。后来我们把这1000平方米拆开,即地面600平方米左右,屋顶300平方米左右。等到开幕期间参加了几次活动以后我就理解了沙特人的真正意图,实际上他们是利用这样一个机会来开展世博外交,也可以称作世博商务或世博公关。他们在这里接待各国元首、举办签字仪式、签署商务合同等,非常气派,既展示了国家文化,还促进了商务外交,以及组织汶川地震的孤儿专场参观等公益活动,充分拓展了世博建筑的功能范畴,把中国提供给他们的这块空间进行了的价值最大化利用。

最后让我用盖里的一段话来结束此次汇报:

"建筑代表的只是人类智慧的一小部分,但是作为从事这个职业的人来说,我深信可以带来改变,可以启迪与丰富人类的体验,可以消除误解,并为生命之歌提供一个美妙的生存场所。"

我非常喜欢这段文字。在此拿来与诸位分享。

Back to the Essentials

Keynote Speech at the 2011 Annual Conference of
the Architectural Society of China

Text / Wang Zhenjun

The Saudi Arabia Pavilion at Expo 2010 has two outstanding features. One is that it is the only venue whose architectural design, including both the interior and landscape, is the sole responsibility of a Chinese team. More importantly, this pavilion has experienced the typical process of being nothing to being something. There was little media coverage of the pavilion before the opening of the Expo. Our team was worried that people would not like it although we had worked on it for two years and a half. After the opening, however, the pavilion became increasingly popular day by day. Visitors might need to wait for nine hours to enter the pavilion when there was a huge crowd.

It is also a process of reflection and learning for architects involved from beginning to end. In particular, it prompts the architects to think about some essential issues about architecture, such as the relationship between form and content, the relationship between form and function, the nature of architecture, the criteria of good architecture, etc.

The Expo is a milestone in the world history of architecture. The process offered me a very rare opportunity to learn and will undoubtedly have major influence on the future of my career. Now I would like to talk about what I have learned in this process and I also want to share with you my views about the real significance of this "beauty pageant of architecture" which I think is the real value of this grand gathering of mankind to architects.

I. Functional Element of National Pavilions at the World Expo

If we look at the 159-year history of the World Expo, it is easy to find that the focus of the Expo changed from the display of technology to the exhibition of culture following the economic crisis of the Western world in the 1960s. To be precise, the event became a platform for countries to show their unique cultures to the world. We can also see at Expo 2010 that the display of technology, or the illustration of technology, is no longer unique to any country. Developed countries may have touchscreen TVs or multimedia. So do developing countries. The focus is now placed on the exhibition and exchange of cultures. In other words, the national pavilions at the World Expo are mainly intended to the display the diverse cultures of participating countries. After some research into the history of the World Expo, I also came to realize that to design a national pavilion is to design a container of culture and what matters most is to put the culture of a country in the container as much as possible. The participation in the design of the Saudi Arabia Pavilion offered me a chance to better understand this nation. I think this is a very smart nation because Saudi Arabia is the only one among the 42 participating foreign countries of Expo 2010 to solicit pavilion design proposals globally. Other countries only invited tenders among domestic architects. I really admire this nation for their courage in this regard. As one of their government officials said, the decision was made right. Some Chinese media outlets have called Saudi Arabia the biggest winner among all foreign countries. I think their courage has been repaid with very good results.

II. Architectural Element of National Pavilions at the World Expo

Three proposals were submitted in the first round of competition in November 2007, namely Moon Boat, Arabian Cube, and Happy Castle. The first two eventually made it to the next round. To develop the design proposal, we needed to gain a deeper understanding of the country's culture, politics, ed ucation, and cities. Therefore, we spent half of less than two weeks studying cultural elements. We also hired Prof. Fu Zhiming from Peking University as our adviser. In retrospect, all the effort we made at the time was worth it. It played a very important role in helping us come up with a clearer design concept or decide on the direction of design.

The concept of Moon Boat draws inspiration from the story of a moon boat drifting from the Middle East to the Shanghai Bay in One Thousand and One Nights. It is also intended to show the long friendship between China and Saudi Arabia. After coming up with the concept, we need to think about how to highlight the element of culture or how to present the culture in a vivid way.

Let's look at the layout of the design. The plot has a rectangle shape but if we change the angle we can see that it points to the direction of Mecca - the holy city to Arabs. The change of axis actually shows respect for Arab culture and gives the pavilion an inte nse religious touch. That was fully recognized by the proprietor. The design of landscape is also based on a thorough study of the Arab culture. Essential elements of the culture are incorporated into the design.

Then we come to the function part. The pavilion is like a museum so it should be built in a way that offers excellent spatial experience.

From the study of the Expo's history we got to know that queuing up for a visit was inevitable. The longest waiting time on record was two hours and a half so we did not expect it to be eight or nine hours. We designed the waiting area to ensure the best warm-up before visit even in the scenario of waiting for two hours and a half.In the waiting area, visitors can queue up around the stage where Arabian dances are performed. This sunshine and temperature graph is very important to our design. The curve suggests that the time from May to October during which the Expo is held is the worst time of the year in Shanghai since it is scorching hot. In light of that, we decided to build a hanging moon boat so as to create a shady and cool waiting area. The second stage comes after the long waiting time and it is an innovation of the entire pavilion. In the beginning, we thought about paving floor slabs. Into the second round of competition, however, we felt the traditional way of visiting a museum would not even intrigue ourselves. The film "Avatar" inspired us a lot. People need to experience things in a more vivid, or more immersed way, with the help of high technology. Old ways, like putting a flat-panel TV, sculpture or robot there, are not enough anymore so we decided to remove the floor slabs. Visitors would feel like hanging in the air and the inner hull of the boat serves as a screen. As one media outlet put it, there are no fixed seats here and you can walk through mountains and rivers on the conveyor to touch the culture of Saudi Arabia. That is exactly the purpose of the design. To put it simply, we just want people to watch films in a different way.

Then we come to the garden on the roof where visitors can have a bird's-eye view of the entire 5.28 square kilometers Expo Site and take great photos. After that, visitors can walk down on a spiral ramp. The shift between indoor and outdoor spaces also gives visitors a stronger impression of the whole process.

III. Element of Green Building

We were purposefully highlighting the concept of green building at the time. It is fair to say that the motivation is to be energy-efficient. It is about integrating the responsibility into the design. We strived to strike a balance between the architectural beauty and environmental friendliness. It is an indisputable fact that it is costly to adopt proper technologies to ensure the maximum use of natural resources.

The energy-saving practices we adopted for the Saudi Arabia Pavilion include the following:

First, the hanging boat design ensures that the pavilion is least impacted by sunshine while no windows on the boat hull reduces heat exchange between the building and the surrounding environment.

Second, considering the hot weather in Shanghai from May to October, we created a shady and comfortable waiting area where performances were presented to the visitors as warm-up activities.

Third, the circular ramps in the pavilion ensure the full use of natural light, thus reducing the reliance on artificial illumination.

Fourth, a patio was built to make full use of natural ventilation and the temperature difference between the upper and lower parts of the building, thus improving air quality in the pavilion.

Fifth, the hallowed outer wall shows an Arab style and also serves as sunshade. The garden on the roof with typical characteristics of Saudi Arabia improves the heat-shielding performance of the roof.

IV. Reflections

First, about the nature of architecture.

More than 2,000 years ago, when Laozi was talking about "Being and Not-being grow out of one another", people already realized that architecture was about making space. The empty space is actually most useful when you are building a house. We were truly amazed by Chinese philosophers' wisdom in designing the Saudi Arabia Pavilion. Space is the key. It should also be noted that people are the key in space since space needs to be experienced by people. This is what Norwegian philosopher Christian Norberg-Schulz called "Genius Loci" - pervading spirit of a place. Only when people and space are both considered can the real architectural experience be created. Some say that architecture has entered an era of experimentation but I think space will always play an essential role in architecture and people is key to space.

Second, about the perception of architecture.

Perception is the degree to which people access information and the architect spreads information. People like the Saudi Arabia Pavilion because it is well perceived. Everybody knows about "One Thousand and One Nights" and the Silk Road. Foreign architects told us they also read the book when they were little. Easily perceptible elements should be included in the design of architecture, particularly Expo and public buildings.

Third, about innovation in architecture.

Design is a reflection of social change. I think this is a very profound insight. People like the interior design of the Saudi Arabia Pavilion perhaps because we responded to the need for high-tech experience in visiting a museum. We offered a different, or more vivid and immersed experience. I often go to museums and they give me a lot of inspiration on this matter. I think architects should deal with different types of architecture but not focus too much on fashions because fashions change and the essentials are eternal.

Fourth, about the social attribute of architecture.

Architects should not seek to design buildings of their personal features. Academician He Jingtang said that all Chinese people are the owners of the China Pavilion. This is talking about the social attribute of architecture, Expo buildings in particular, which cannot be embodied in other art forms. Architecture is an art of space and it has many cultural missions to fulfill. From this perspective, architecture is not simply about the aesthetics of the architect but also about the appreciation of other stakeholders including the general public.

Fifth, good architecture should be realistic.

I often ask myself what is good architecture and the answer I find include five factors. Firstly, true feelings are integrated into the design. Secondly, the concept is clearly perceptible. Thirdly, the form and content are highly consistent. If the space in a building is useless, it is not good architecture. Fourthly, the spatial experience moves people deeply. Fifthly, cultural elements are well imbedded in the design given that architecture has a role to play in the promotion of culture.

Sixth, about the bigger role of Expo buildings.

In previous Expos, the VIP area of a national pavilion would not exceed 100 squaremeters but Saudi Arabia wanted to be 1,000 squaremeters this time. I could not understand it at first because that would undermine the overall look of the pavilion. What we did was to create a 600 square meters VIP area on the ground and a 300 square meters on the roof. I came to understand it after attending activities on several occasions. The Saudi Arabians are so smart. They were actually taking the opportunity to exercise diplomacy. Many business contracts were signed there; signing ceremonies were held there; many heads of state were received there. They were making full use of the plot offered by Shanghai to display their culture and organize business activities. The organized visit to the pavilion by children orphaned by the Wenchuan earthquake was a very good activity to establish a charitable image.

The great architect Frank Gehry once said, "Architecture is a small piece of this human equation, but for those of us who practice it, we believe in its potential to make a difference, to enlighten and to enrich the human experience, to penetrate the barriers of misunderstanding and provide a beautiful context for life's drama."

This is a quote I love very much and I think every architect should bear that in mind.

作品年表
Chronology of Works

项目名称	北京新世界中心
建造地点	北京
设计时间	1993 年
竣工时间	1998 年
建设规模	19.9 万平方米
功能类型	商业、酒店公寓综合体
业主	北京崇文新世界房地产发展有限公司

项目名称	三亚山海天大酒店
建造地点	海南三亚
设计时间	1995 年
竣工时间	1998 年
建设规模	2.72 万平方米
功能类型	酒店
业主	山东鲁能房地产有限公司

项目名称	上海浦东软件园一期、二期研发建筑群
建造地点	上海
设计时间	1995 年
竣工时间	2003 年
建设规模	15.6 万平方米
功能类型	科技园区
业主	上海浦东软件园有限责任公司

项目名称	首都国际机场新塔台
建造地点	北京
设计时间	2000 年
竣工时间	2002 年
建设规模	0.56 万平方米
功能类型	空中交通控制塔
业主	中国民航华北空管局

项目名称	北京亚运村汽车交易新市场
建造地点	北京
设计时间	2002 年
竣工时间	2004 年
建设规模	7.54 万平方米
功能类型	展览、商业
业主	北京北辰集团公司

项目名称	北京国际财源中心
建造地点	北京
设计时间	2004 年
竣工时间	2010 年
建设规模	24.42 万平方米
功能类型	办公、商业综合体
业主	北京建机天润房地产有限公司

项目名称	首都国际机场东区塔台
建造地点	北京
设计时间	2004 年
竣工时间	2007 年
建设规模	0.28 万平方米
功能类型	空中交通控制塔
业主	中国民航华北空管局

项目名称	上海国家软件出口基地三期
建造地点	上海
设计时间	2005 年
竣工时间	2012 年
建设规模	44.77 万平方米
功能类型	科技园区
业主	上海浦东软件园有限责任公司

项目名称	南京江东国际外包服务基地规划
建造地点	南京
设计时间	2005 年
建设规模	用地 110.99 公顷
功能类型	科技园区
业主	南京鼓楼国际软件与服务外包产业园有限公司

项目名称	昆山浦东软件园
建造地点	江苏苏州
设计时间	2006 年
竣工时间	2009 年
建设规模	44.25 万平方米
功能类型	科技园区
业主	上海浦东软件园有限公司

项目名称	2010 上海世博会沙特国家馆
建造地点	上海
设计时间	2007 年
竣工时间	2010 年
建设规模	0.61 万平方米
功能类型	博览建筑
业主	沙特阿拉伯城乡事务部

项目名称	上海虹桥国际机场新塔台及附属建筑
建造地点	上海
设计时间	2007 年
建设规模	0.57 万平方米
功能类型	空中交通控制塔
业主	中国民航华东空管局

项目名称	天津软通动力软件园总体规划
建造地点	天津
设计时间	2008 年
建设规模	用地 20.11 公顷
功能类型	科技园区
业主	软通动力信息技术有限公司

项目名称	嘉兴高新软件园规划
建造地点	浙江嘉兴
设计时间	2008 年
建设规模	用地 89.77 万公顷
功能类型	科技园区
业主	嘉兴高新软件园有限责任公司

项目名称	长沙中电软件园总部大楼及配套工程
建造地点	湖南长沙
设计时间	2009 年
竣工时间	2013 年
建设规模	5.76 万平方米
功能类型	综合体，智能建筑
业主	长沙中电软件园有限公司

项目名称	中国建设银行股份有限公司北京生产基地
建造地点	北京
设计时间	2009 年
建设规模	28.11 万平方米
功能类型	产业园区
业主	中国建设银行股份有限公司

项目名称	西安咸阳国际机场新塔台及附属建筑
建造地点	陕西西安
设计时间	2009 年
竣工时间	2012 年
建设规模	0.56 万平方米
功能类型	空中交通控制塔
业主	中国民航西北空管局

项目名称	苏州水墨江南
建造地点	江苏苏州
设计时间	2009 年
竣工时间	2011 年
建设规模	12 万平方米
功能类型	酒店、居住建筑
业主	苏州工业园区建屋物业发展有限公司

项目名称 天津滨海国际机场第二跑道空管工程航管楼塔台设计
建造地点 天津
设计时间 2010年
竣工时间 2014年
建设规模 0.61万平方米
功能类型 空中交通控制塔
业主 中国民航华北空管局天津分局

项目名称 北京泰德制药股份有限公司二期扩建项目
建造地点 北京
设计时间 2010年
竣工时间 2014年
建设规模 5.81万平方米
功能类型 综合体，智能建筑
业主 北京泰德制药股份有限公司

项目名称 中国国学中心国际竞赛
建造地点 北京
设计时间 2010年
建设规模 5.98万平方米
功能类型 博览建筑
业主 国务院参事室

项目名称 固安城市规划展馆
建造地点 河北固安
设计时间 2010年
竣工时间 2014年
建设规模 1.01万平方米
功能类型 博览建筑
业主 华贸幸福基业资产管理有限公司

项目名称 中国证券期货业信息服务基地
建造地点 北京
设计时间 2011年
建设规模 用地27.12公顷
功能类型 科技园区
业主 中国证券登记结算有限责任公司

项目名称 上海电影博物馆暨电影艺术研究所业务大楼项目
建造地点 上海
设计时间 2011年
建设规模 3.67万平方米
功能类型 展览，办公建筑
业主 上海电影艺术研究所

项目名称 美国塞班岛别墅区
建造地点 美国塞班岛
设计时间 2011年
竣工时间 2017年
建设规模 1.76万平方米
功能类型 居住建筑
业主 领鑫国际投资有限公司

项目名称 中国智慧谷房山数据中心
建造地点 北京
设计时间 2012年
建设规模 3.98万平方米
功能类型 数据研发中心
业主 北京能通科技股份有限公司

项目名称 中华回乡文化园沙特国家馆姊妹馆
建造地点 宁夏回族自治区银川市
设计时间 2012年
竣工时间 2018年
建设规模 0.79万平方米
功能类型 博览建筑
业主 宁夏回乡文化实业有限公司

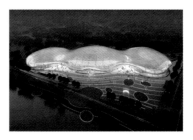

项目名称 浙江永嘉体育中心竞赛
建造地点 浙江温州
设计时间 2012年
建设规模 1.86万平方米
功能类型 体育建筑
业主 永嘉上塘中心城区管理委员会

项目名称 中国信达（合肥）灾备及后援基地
建造地点 安徽合肥
设计时间 2012年
竣工时间 2016年
建设规模 13.98万平方米
功能类型 数据灾备建筑
业主 中国信达资产管理股份有限公司安徽省分公司
合作设计 世源科技工程有限公司

项目名称 百度阳泉云计算中心
建造地点 山西阳泉
设计时间 2012年
建设规模 14.67万平方米
功能类型 科技园区
业主 百度在线网络技术（北京）有限公司

项目名称　吴都阖闾城遗址博物馆多媒体展厅
建造地点　江苏无锡
设计时间　2012 年
竣工时间　2014 年
建设规模　0.06 万平方米
功能类型　博览建筑
业主　　　无锡阖闾城博物馆建设有限公司

项目名称　中国建设银行党校教学用房
建造地点　北京
设计时间　2012 年
建设规模　0.70 万平方米
功能类型　教育建筑
业主　　　中国建设银行股份有限公司

项目名称　上海智慧岛数据产业园办公园区城市设计
建造地点　上海
设计时间　2012 年
建设规模　用地 218.11 公顷
功能类型　科技园区
业主　　　崇明县智慧岛开发建设有限公司

项目名称　鄂尔多斯云计算中心规划
建造地点　内蒙古鄂尔多斯
设计时间　2012 年
建设规模　用地 493.3 公顷
功能类型　科技园区
业主　　　鄂尔多斯高新区管委会

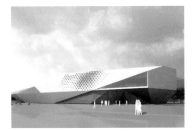

项目名称　郑州航空经济试验区规划展示中心
建造地点　河南郑州
设计时间　2012 年
建设规模　1.42 万平方米
功能类型　博览建筑
业主　　　郑州航空港建设局

项目名称　燕郊世界华人收藏博物馆
建造地点　河北三河
设计时间　2013 年
竣工时间　2018 年
建设规模　4.93 万平方米
功能类型　博览建筑、酒店
业主　　　百世佳联房地产开发有限公司

项目名称　山西三元煤业股份有限公司技术研发中心项目
建造地点　山西长治
设计时间　2013 年
建设规模　31.87 万平方米
功能类型　科技园区
业主　　　山西三元煤业股份有限公司

项目名称　中国瑞达投资发展集团公司瑞达石景山路 23 号院科研基地
建造地点　北京
设计时间　2013 年
建设规模　14.82 万平方米
功能类型　综合体、智能建筑
业主　　　中国瑞达投资发展集团公司

项目名称　中央美术学院燕郊校区教学楼
建造地点　河北三河
设计时间　2013 年
竣工时间　2015 年
建设规模　3.43 万平方米
功能类型　教育建筑
业主　　　中央美术学院

项目名称　中央美术学院燕郊校区风雨操场建筑设计
建造地点　河北三河
设计时间　2013 年
竣工时间　201 年
建设规模　0.19 万平方米
功能类型　体育建筑
业主　　　中央美术学院

项目名称　潞安·中央公元住宅区设计
建造地点　山西临汾
设计时间　2013 年
建设规模　69.2 万平方米
功能类型　居住建筑
业主　　　潞安鸿源房地产开发有限公司

项目名称　郑州新郑国际机场新塔台及附属建筑
建造地点　河南郑州
设计时间　2013 年
竣工时间　2016 年
建设规模　1.32 万平方米
功能类型　空中交通控制塔
业主　　　中国民航中南空管局

项目名称	郑州航空港银河办事处第二邻里中心小学
建造地点	河南郑州
设计时间	2014 年
建设规模	1.32 万平方米
功能类型	教育建筑
业主	郑州航空港规划局

项目名称	天津逸仙科学工业园转型概念规划
建造地点	天津武清
设计时间	2014 年
建设规模	用地 289 公顷
功能类型	产业园区
业主	天津经济技术开发区逸仙科学工业园

项目名称	海盐杭州湾智能制造创新中心
建造地点	浙江嘉兴
设计时间	2014 年
竣工时间	2018 年
建设规模	27.5 万平方米
功能类型	科技园区
业主	海盐经济开发区管理委员会

项目名称	中关村移动智能创新服务园
建造地点	北京
设计时间	2014 年
竣工时间	2019 年
建设规模	6.98 万平方米
功能类型	科技园区
业主	北京首农信息产业投资有限公司

项目名称	河南鄢陵康泰半岛花语别墅区
建造地点	河南鄢陵县
设计时间	2014 年
建设规模	46.76 万平方米
功能类型	居住建筑
业主	河南康泰置业有限公司

项目名称	阿克苏三馆及市民服务中心
建造地点	新疆阿克苏
设计时间	2015 年
建设规模	5.98 万平方米
功能类型	博览建筑
业主	新疆阿克苏规划局

项目名称	北京新机场空防安保培训中心
建造地点	北京
设计时间	2015 年
建设规模	3.0 万平方米
功能类型	综合体、智能建筑
业主	北京新机场建设指挥部

项目名称	北京新机场信息中心 & 指挥中心
建造地点	北京
设计时间	2015 年
竣工时间	2019 年
建设规模	2.99 万平方米
功能类型	数据中心、智能建筑
业主	北京新机场建设指挥部

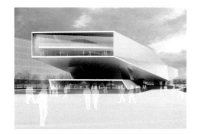

项目名称	青岛惠普全球大数据应用研究及产业示范基地展示中心
建造地点	山东青岛
设计时间	2015 年
建设规模	4.42 万平方米
功能类型	博览建筑
业主	青岛博康国际发展有限公司

项目名称	中国人民银行反洗钱上海监测分析中心
建造地点	上海
设计时间	2015 年
建设规模	2.82 万平方米
功能类型	办公、数据中心、智能建筑
业主	中国人民银行

项目名称	郑州阿里云中部创业创新基地
建造地点	河南郑州
设计时间	2016 年
建设规模	28.12 万平方米
功能类型	科技园区
业主	河南中原云港发展有限公司

项目名称	潍坊崇文新农村综合服务基地
建造地点	山东潍坊
设计时间	2016 年
竣工时间	2018 年
建设规模	20.1 万平方米
功能类型	科技园区
业主	潍坊恒建资产经营管理有限公司

项目名称	湖北三峡移民博物馆
建造地点	湖北秭归
设计时间	2016年
建设规模	1.0万平方米
功能类型	博览建筑
业主	湖北宜昌秭归县规划局

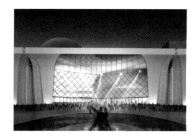

项目名称	塔城市文化艺术中心
建造地点	新疆塔城
设计时间	2016年
建设规模	1.6万平方米
功能类型	文化建筑
业主	塔城市致远文化发展有限公司

项目名称	北京新机场行政综合业务用房工程
建造地点	北京
设计时间	2016年
竣工时间	2019年
建设规模	6.9万平方米
功能类型	综合体、智能建筑
业主	北京新机场建设指挥部

项目名称	北京新机场生活服务设施工程
建造地点	北京
设计时间	2016年
竣工时间	2019年
建设规模	6.6万平方米
功能类型	综合体、智能建筑
业主	北京新机场建设指挥部

项目名称	北京新机场工作区物业工程
建造地点	北京
设计时间	2016年
竣工时间	2019年
建设规模	2.1万平方米
功能类型	综合体、智能建筑
业主	北京新机场建设指挥部

项目名称	北京新机场工作区车辆维修中心工程
建造地点	北京
设计时间	2016年
竣工时间	2019年
建设规模	0.99万平方米
功能类型	综合体、智能建筑
业主	北京新机场建设指挥部

项目名称	北京按摩医院
建造地点	北京
设计时间	2016年
竣工时间	2019年
建设规模	3.89万平方米
功能类型	医院
业主	北京按摩医院

项目名称	北京雁栖湖山地艺术家工作区
建造地点	北京
设计时间	2016年
建设规模	11.9万平方米
功能类型	居住建筑
业主	江河创新地产股份有限公司

项目名称	北京新机场东方航空公司综合业务楼、运行楼及宿舍
建造地点	北京
设计时间	2017年
建设规模	40.86万平方米
功能类型	办公综合体、宿舍
业主	中国东方航空公司

项目名称	非主基地航空公司办公及宿舍
建造地点	北京
设计时间	2017年
竣工时间	2019年
建设规模	25.97万平方米
功能类型	办公综合体、宿舍
业主	北京首地兴业置业有限公司

项目名称	河南省安阳市豫东北机场新建航站楼及塔台概念性方案
建造地点	河南安阳
设计时间	2017年
建设规模	2.1万平方米
功能类型	航站楼
业主	安阳市交通机场发展投资有限责任公司

项目名称	芜湖皖南医学院弋矶山医院三山医养结合示范园区
建造地点	安徽芜湖
设计时间	2017年
建设规模	29.82万平方米
功能类型	医院，养老建筑
业主	芜湖金晖三华健康产业投资有限公司

作品获奖
Awards

作品 1：湘西永顺县王村镇规划
  荣获建设部 1989 年度优秀规划设计二等奖
作品 2：郑州市科技馆与博物馆方案设计
  荣获全国八大高校参加的设计竞赛首奖
作品 3：北京新世界中心一期工程
  荣获北京九十年代十大建筑称号；
  中国建筑学会建筑创作大奖入围奖
作品 4：三亚山海天大酒店
  荣获海南省 2001 年度优秀设计一等奖；
  建设部优秀设计三等奖；
  （该作品刊登在《Architecture China》杂志第一期）
作品 5：上海浦东软件园一、二期工程
  荣获信息产业部 2000、2004 年度优秀设计一等奖；
  中国建筑学会建筑创作大奖入围奖
作品 6：首都机场新塔台
  荣获信息产业部 2006 年度优秀设计二等奖
作品 7：上海国家软件出口基地
  （浦东软件园三期）
  荣获国际竞赛首奖；
  工信部 2010 年度优秀设计一等奖
作品 8：德国 SAP 中国研究院
  荣获信息产业部 2008 年度优秀设计二等奖
作品 9：首都机场东区塔台
  荣获国家优秀工程设计银质奖；
  中国建筑学会六十周年建筑创作大奖；
  工业和信息化部 2008 年度优秀设计一等奖
作品 10：2010 年上海世博会沙特国家馆
  荣获国际竞赛首奖；
  2010 年上海世博会创意展示 A 类金奖 (国际展览局 BIE 颁)；
  第六届中国建筑学会建筑创作金奖；
  第六届中国威海国际建筑特别奖；
  中国勘察设计协会 2011 年度行业优秀建筑工程设计二等奖；
  中国勘察设计协会 2011 年度行业优秀智能化建筑一等奖；
  腾讯网上海世博会十大最佳展馆奖；
  工信部 2010 年度优秀设计一等奖；
  第十八届首都城市规划建筑设计方案汇报展优秀方案奖
作品 11：长沙中电软件园总部大楼
  荣获中国建筑学会 2013 年度中国建筑设计奖银奖；
  北京第十七届优秀工程设计二等奖
作品 12：北京国际财源中心（IFC）
  荣获中国勘察设计协会 2013 年度建筑设计行业奖一等奖；
  首都第二十一届城市规划建筑设计（方案）展一等奖
作品 13：北京泰德制药研发中心扩建项目
  荣获北京市 2014 年度优秀建筑设计二等奖；
  荣获中国勘察设计协会 2015 年度建筑设计行业三等奖
作品 14：中国信达（合肥）灾备及后援基地
  荣获中国勘察设计协会 2017 年度建筑设计行业二等奖

Work 1: The plan of Wangcun Town, Yongshun County, Hunan
  Second Prize of Excellent Design Prize Awarded by the Ministry of Construction
Work 2: Concept design for Zhengzhou Science and Technology Museum and Museum
  First Prize of Design Competition Attended by Eight Chinese Universities
Work 3: Beijing New World Center (Phase I)
  Beijing Top 10 Buildings in 1990s;
  Nomination-prize of ASC Architectural Creation Award
Work 4: Sanya Shanhaitian Hotel
  First Prize of Hainan Excellent Design Award 2001;
  Third Prize of Ministry of Construction Excellent Design Award
  (the works is published in Architecture China, issue 1)
Work 5: Shanghai Pudong Software Park (Phase I / II)
  First Prize of Ministry of Information Industry Excellent Design Award in 2000 and 2004;
  Nomination-prize of ASC Architectural Creation Award
Work 6: The new control tower for Capital Airport
  Special Prize of Ministry of Information Industry Excellent Design Award in 2006
Work 7: The conceptual planning design for Shanghai National Software Export Base
  (Pudong Software Park Phase III)
  First Prize in International Competition;
  First Prize of the Ministry of Industry and Information Technology Excellent Design Award in 2010
Work 8: SAP China Research Institute
  Second Prize of Ministry of Information Industry Excellent Design Award in 2008
Work 9: The control tower for east zone, Capital Airport
  National Excellent Engineering & Exploration Design Silver Award;
  ASC Architectural Creation Award;
  First Prize of Ministry of Industry and Information Technology Excellent Design Award in 2008
Work 10: The Saudi Arabia Pavilion at Shanghai World Expo 2010
  First Prize in International Competition;
  Creative Exhibition Category A Gold Prize for Shanghai World Expo 2010 (awarded by BIE);
  The Sixth ASC Architectural Creation Award;
  Special Prize of The Sixth Weihai International Architecture Award;
  Second Prize of China Exploration & Design Association Excellent Architectural Design Award in 2011;
  First Prize of China Exploration & Design Association Excellent Intelligent Building Award in 2011;
  Tencent.com Top Ten Pavilions at Shanghai World Expo 2010;
  First Prize of Ministry of Industry and Information Technology Excellent Design Award in 2010;
  The 18th Capital Urban Planning Architectural Design Schemes Exhibition Excellent Scheme Prize
Work 11: The Headquarters Building for Changsha CEC Software Park
  Silver Prize of ASC China Architectural Design Award in 2013;
  Second Prize of the 17th Beijing Excellent Architectural Design Award
Work 12: Beijing IFC
  First Prize of China Exploration & Design Association Architectural Design Award in 2013;
  First Prize of the 21st Capital Urban Planning Architectural Design (Schemes)
Work 13: The Expansion of Beijing TIDE Pharmaceutical R&D Center
  Second Prize of the 18th Beijing Excellent Architectural Design Award;
  Third Prize of China Exploration & Design Association Excellent Architectural Design Award in 2015
Work 14: China Cinda (Hefei) Disaster Recovery and Back-up Base
  Second Prize of China Exploration & Design Association Excellent Architectural Design Award in 2017

媒体评价
Reviews

- 王振军设计作品及设计理念被分别刊登在由中国建筑学会主编的《中国青年建筑师1》及中国建筑学会建筑师分会《建筑创作》杂志社主编的《中国青年建筑师188》和《中国青年建筑师·当代中国新作品——创作者自画像》书中。

- 因作品荣获"北京九十年代十大建筑"奖，王振军作为主要设计人接受《首都建设报》记者蔡青采访，采访文章发表在2001年7月20日《首都建设报》第二版 "现代中叠现传统神韵"。

- 因作品荣获"上海国家软件出口基地（浦东软件园三期）规划设计" 竞赛首奖，王振军作为总设计师接受上海《新民晚报》记者吴强采访，采访文章发表在2001年4月5日《新民晚报》第十九版。

- 因荣获"2010上海世博会沙特国家馆"国际竞赛首奖，王振军作为总设计师接受《上海世博》记者王焱冰专访，专访录发表在《上海世博》2009年第12期，中国建筑的核心期刊《建筑学报》2008年12期、《世界建筑》2009年1期、《时代建筑》2009年1期，以及《建筑观察》2009年1期均相继对此进行了专题报道。2010年4月~10月接受《中华建筑报》《中国建设报》《人民日报》《新民周刊》《瞭望东方周刊》等专访。

- 2010年4月接受中国中央电视台（CCTV）大型纪录片《世博的记忆》和《绿色建筑》栏目专访。

- 2010年5月~8月分别接受凤凰网、新浪网、腾讯网等媒体采访。

- 2010年9月23日接受CCTV新闻频道"天天世博会"栏目专访。

- 2011年5月"王振军访谈录"发表在《甲方乙方》。

- 2011年9月23日最新作品被刊登在中国建筑学会主编的《前进中的中国建筑1993~2010》，并在日本召开的世界建协大会期间展出。

- 因沙特馆在上海世博会上的突出表现被原地保留，并于2011年9月率先重新对外开放，王振军接受搜狐网采访。

- 2012年1月"王振军访谈录"发表在《设计过程》。

- 2012年1月在由中国建筑学会举办的"繁荣和发展中国建筑文化座谈会"（人民大会堂）发言"建筑的本质与中国建筑师的文化使命"。

- 2013年3月接受河南卫视"根在中原"节目专访，播出题目：城市造梦者。

- 2013年7月"商务办公建筑的两极化发展"载于2013.7.18《中国建设报》。

- 2014年9月最新作品被刊登在中国建筑学会主编的《全球化进程中的当代中国建筑》。

- The design works and design concept of Wang Zhenjun are published in "Chinese Young Architects I" compiled by ASC, and "Chinese Young Architects 188" and "Self-portrait for Creators: Chinese Young Architects & New Architectural Works in Contemporary China" compiled by the journal press of Architectural Creation, run by the Architects Sub-society of ASC.

- After his work being awarded Beijing Top 10 Buildings in 1990s, Wang Zhenjun, as the chief designer, was interviewed by Cai Qing, a reporter with Capital Construction News, and the report ——"Traditional Verve Mixed in Modern Characteristics" was published on page 2 of Capital Construction News on July 20, 2001.

- As his works won the first prize in the competition of "Planning Design for Shanghai National Software Export Base (Pudong Software Park Phase III)", Wang Zhenjun, as the chief architect, was interviewed by Wu Qiang, a correspondent with Xinmin Evening News, and the interview coverage was published on page 19 of Xinmin Evening News dated April 5, 2001.

- After winning the first prize of the International Competition for Saudi Arabia Pavilion at Shanghai World Expo in 2010, Wang Zhenjun, as the Chief Architect, was interviewed by Wang Yanbing, a journalist with Shanghai World Expo, and the interview record was published in Shanghai World Expo, issue 12 of 2009; it was reported by a series of core journals dedicated to architecture in China, such as Architectural Journal, issue 12 of 2008; World Architecture, issue 1 of 2009; Time Architecture, issue 1 of 2009 and Architecture View, issue 1 of 2009. He was also interviewed by China Construction News, China Construction Gazette, People's Daily, Xinmin Weekly, and Oriental Outlook from April through October, 2010.

- Interviewed by World Expo Memory and Green Architecture, documentary channels of CCTV in April, 2010.

- Interviewed by sina.com, ifeng.com and tecent.com in May ~ August, 2010.

- Interviewed by World Expo Everyday of the CCTV News channel on September 23, 2010.

- "Interview Record of Wang Zhenjun", Party A and Party B, May, 2011.

- His latest works are included in "Perspective of Chinese Architecture 1993 ~ 2010". Selected Works of ASC Young Architects Awards compiled by Architectural Society of China on September 23, 2011, and they were on display during the 24th UIA World Congress of Architecture in Tokyo.

- The Saudi Arabia Pavilion is retained at the original site due to its outstanding performance during the Shanghai World Expo, and it is open to public again in September, 2011. Interviewed by sohu.com.

- "Interview Record of Wang Zhenjun", Design Process, January, 2012.

- "The Essence of Architecture and the Cultural Mission of Chinese Architects", a speech on the symposium on flourishing and developing Chinese architectural culture (held in the Great Hall of the People), January, 2012.

- Interviewed by "Rooted in Central China" program of Henan Satellite TV in March 2013, and the interview was broadcast in the title of Urban Dream-maker.

- "The Polarized Development of Business Architecture", China Construction Gazette, July 18, 2013.

- In September 2014, his latest design works were published in "Contemporary Chinese Architecture in the Globalization Process" compiled by ASC.

参展
Exhibition

- 2009 年 10 月 2010 上海世博会建筑设计作品展，上海
- 2011 年 5 月 北京中国建筑学会建筑创作奖作品展，北京
- 2011 年 9 月 东京世界建协大会"前进中的中国建筑 1993~2010 中国建筑作品展 "，日本东京
- 2012 年 10 月 北京中国建筑学会"中国当代百名建筑师"作品展，北京
- 2013 年 7 月 第八届威海国际建筑节"中国当代百名建筑师"作品展，威海
- 2014 年 9 月 UIA 南非德班第 25 届世界建筑师大会"全球化进程中的当代中国建筑"作品展，南非德班
- 2016 年 10 月 韩国 2016 年"百名建筑师"展览，韩国

- October 2009, Shanghai World Expo 2010 Architectural Design Works Exhibition, Shanghai
- May 2011, Beijing ASC Architectural Creation Award Works Exhibition, Beijing
- September 2011, Perspective of Chinese Architecture 1993 ~ 2010: Chinese Architectural Works Exhibition during the 24th UIA World Congress of Architecture, Tokyo
- October 2012, Beijing ASC Exhibition of Works of "100 Contemporary Chinese Architects", Beijing
- July 2013, Exhibition of Works of "100 Contemporary Chinese Architects"during the 8th Weihai International Architecture Festival, Weihai
- September 2014, "Contemporary Chinese Architecture in the Globalization Process"—— Chinese Architectural Works Exhibition during the 25th UIA World Congress of Architecture, Durban, South Africa.
- 2016 International Exhibition of "100 Architects of the Year" , Korea

论著
Treatises

- 参加第一届在湖南长沙举办的全国"建筑与文化"研讨会并在会上宣读论文。
- 参加第 93 次青年科学家论坛"当代建筑——关于技术的挑战"并在会上宣读论文。
- 参加 1998 年青年建筑师学术讨论会并在会上宣读论文。

- Attended the first national seminar on "Architecture and Culture" and read a dissertation on the seminar held in Changsha, Hunan.
- Attended the 93rd Young Scientists Forum "Contemporary Architecture ── On the Challenge of Technologies" and read a dissertation on the forum.
- Attended 1998 Young Architects Symposium and read a dissertation on the symposium.

国家核心及主要期刊发表论文和演讲

论文 1：《思维结构的调整与建筑文化的新图景》
　　　　1990 年第 2 期《南方建筑》；
论文 2：《精神与形式——关于安藤忠雄与后现代主义》
　　　　1991 年第 1 期《时代建筑》；
论文 3：《轴线手法与建筑空间及形式的层次感塑造》
　　　　1998 年第 5 期《工程建设与设计》；
论文 4：《人与自然通过科技在信息时代的整合
　　　　——上海浦东软件园设计》
　　　　2000 年第 11 期《建筑学报》；
论文 5：《探索软件园设计的生态学途径
　　　　——上海国家软件出口基地规划设计》
　　　　2005 年第 1 期《建筑学报》；
论文 6：《"山海天大酒店"工程介绍》
　　　　2005 年第 1 期《名筑》；
论文 7：《建筑秀场上的文化容器》
　　　　2010 年第 5 期《建筑学报》；
论文 8：《塑造阿拉伯文化的诗意体验》
　　　　2010 上海世博会城市最佳实践区暨国际建筑论坛上发表的演讲；
论文 9：《传统技艺，诗意体验——沙特馆》
　　　　2010 年第 9 期《建筑技艺》；
论文 10：《文化容器里的诗意体验》
　　　　2010 年第 10 期《建筑与文化》；
论文 11：《回归本质》
　　　　2011 年中国建筑学会学术论坛上发表的主旨演讲；
论文 12：《小舟撑起绿茵来——卡米诺·努埃奥特许学校》
　　　　2011 年第 1 期《建筑与文化》；
论文 13：《城市消失》
　　　　2011 年第 11 期《建筑与文化》；

The Essays and Speeches Published on National Core and Primary Journals

Essay　1: Adjustment of Thinking Structure and the New Vision of Architectural Culture
　　　　South Architecture, No. 02, 1990
Essay　2: Spirit and form ── on Ando Tadao and Postmodernism
　　　　Time Architecture, No. 01, 1991
Essay　3: The Axis Technique and the Shaping of Layering in Architectural Space and Form
　　　　Construction & Design for Project, No. 05, 1998
Essay　4: Integration of Man and Nature in Information Age via Technology
　　　　── The Design of Shanghai Pudong Software Park
　　　　Architectural Journal,No. 11, 2000
Essay　5: Exploring the Ecological Approach for the Design of Software Parks
　　　　── Planning Design for Shanghai National Software Export Base
　　　　Architectural Journal, No. 01, 2005
Essay　6: Introduction of Sanya Shanhaitian Hotel
　　　　Famous buildings, No. 01, 2005
Essay　7: Cultural Container on the Arena for Architectural Show
　　　　Architectural Journal, No. 05, 2010
Essay　8: Shaping the Poetic Experience of Arab Culture
　　　　Speech on 2010 Shanghai World Expo Urban Best Practice Area & International Architecture Forum
Essay　9: Traditional Technique,Poetic Experience,Saudi Arabia Pavilion
　　　　Architecture Technique, No. 09, 2010
Essay 10: The Poetic Experience of Culture Container
　　　　Architecture & Culture, No. 10, 2010
Essay 11: Return to nature
　　　　The Keynote Speech on Academic Forum of ASC, 2011
Essay 12: A Boat Brings A Shade ── Camino Nuevo High School, Los Angeles
　　　　Architecture & Culture, No. 01, 2011
Essay 13: City Disappear
　　　　Architecture & Culture, No. 11, 2011

论文 14：《建筑的本质与中国建筑师的文化使命》
　　　　2012 年第 2 期《建筑学报》；
　　　　中国建筑学会人民大会堂组织的研讨会上宣读；
论文 15：《对折板——法国博索莱伊小学浅析》
　　　　2012 年第 3 期《建筑与文化》；
论文 16：《记忆碎片——西班牙复活节博物馆浅析》
　　　　2012 年第 7 期《建筑与文化》；
论文 17：《速度与激情在空间上的平衡
　　　　——保时捷汽车博物馆》
　　　　2012 年第 11 期《建筑与文化》；
论文 18：《质疑的时代更需要标准》
　　　　2013 年第 2 期《建筑与文化》；
论文 19：《商务建筑的两级化发展》
　　　　2013 年 7 月 18 日《中国建设报》；
论文 20：《文化之塔——中国国学中心》
　　　　2013 年第 2 期《建筑与文化》；
论文 21：《建筑介入自然》
　　　　2014 年第 8 期《建筑与文化》；
论文 22：《海港上的水晶——NYKREDIT 新总部大楼设计策略浅析》
　　　　2015 年第 3 期《建筑与文化》；
论文 23：《中国瑞达石景山路 23 号院科研基地》
　　　　2015 年第 4 期《建筑与文化》；
论文 24：《郑州航空经济实验区规划展馆设计》
　　　　2015 年第 6 期《建筑与文化》；
论文 25：《PTTEP——S1 办公楼项目设计品鉴》
　　　　2015 年第 12 期《建筑与文化》；
论文 26：《高科技园区规划设计和建设的思考》
　　　　2017 年第 10 期《建筑技艺》；

CEEDI 院内期刊

论文 1：《五维空间——试析 M · 波塔的一件室内设计作品》
　　　　1996 年第 10 期《设计与信息》；
论文 2：《重组的诱惑——民航华北空管局办公楼创作体验》
　　　　1997 年第 4 期《设计与信息》；
论文 3：《建筑形式与精神场所——首都机场新塔台设计有感》
　　　　2000 年第 4 期《设计与信息》；
论文 4：《上海国家软件出口基地（浦东软件园三期工程）规划方案设计》
　　　　2004 年第 2 期《设计与信息》；

著作

著作 1：《轴线手法在当代建筑设计中的应用》
　　　　中国建筑工业出版社，2017 年 2 月出版；
著作 2：《蔓设计》
　　　　中国建筑工业出版社，2017 年 10 月出版；

技术标准、规范编制

规范 1：《软件园区规划设计规范》主编
　　　　2013 年 10 月颁布发行；
规范 2：《机场指挥塔台建筑设计规范》主编，编制中；
规范 3：《电子信息系统机房设计规范》编委，编制中；
规范 4：《机场航站楼室内装饰装修工程技术规程》编委，编制中

Essay 14: The Essence of Architecture and The Cultural Mission of Chinese Architects
　　　　Architectural Journal, No. 02, 2012
Essay 15: Folded Plate —— Analysis on Primary School in Beausoleil by Calori Azimi Botineau Architects in Beausoleil, France
　　　　Architecture & Culture, No. 03, 2012
Essay 16: Fragments of Memories—— Analysis on Easter Sculpture Museum
　　　　Architecture & Culture, No. 07, 2012
Essay 17: Space Balance between Speed and Emotion
　　　　—— Porsche Museum Designed by Delugan Meissl Associated Architects
　　　　Architecture & Culture, No. 11, 2012
Essay 18: The Necessity of Standard in Oppugning Time
　　　　Architecture & Culture, No. 02, 2013
Essay 19: The Polarized Development of Business Architecture
　　　　China Construction News, 07, 18, 2013
Essay 20: A Tower of Civilization—— China Sinology Center
　　　　Architecture & Culture, No. 02, 2013
Essay 21: Architecture in nature
　　　　Architecture & Culture, No. 08, 2014
Essay 22: Harbour Crystal—— Analysis of Design Strategies on The New Headquarters Building
　　　　Architecture & Culture, No. 03, 2015
Essay 23: Reida Shijingshan Road No. 23 Hospital Research Base
　　　　Architecture & Culture, No. 04, 2015
Essay 24: The Design of Exhibition Hall in Zhengzhou Aviation Economy Area
　　　　Architecture & Culture, No. 06, 2015
Essay 25: PTTEP-S1 Office
　　　　Architecture & Culture, No. 12, 2015
Essay 26: Thoughts on Planning, Design and Construction of High Tech Parks
　　　　Architecture Technique, No. 10, 2017

CEEDI Internal Periodical

Essay 1: 5D Space —— An Analysis of Mario Botta's Interior Design Work
　　　　Design and Information, No. 10, 1996
Essay 2: The Temptation of Restructuring
　　　　—— The Experience of Design for CAAC North China Air Traffic Control Bureau Office Building
　　　　Design and Information, No. 04, 1997
Essay 3: Architectural Form and Spiritual Place
　　　　—— Thoughts on The Design of New Control Tower for Capital Airport
　　　　Design and Information, No. 04, 2000
Essay 4: Planning Design for Shanghai National Software Export Base (Pudong Software Park Phase III)
　　　　Design and Information, No. 02, 2004

Books

Book 1: The Application of Axis Approach in Contemporary Architectural Design
　　　　China Architecture & Building Press, 2017.02
Book 2: Organically-Permeated Design
　　　　China Architecture & Building Press, 2017.10

Codes

Code 1: Code of Software Park Planning & Design
　　　　Date Issued: 10/2013
Code 2: Code for Architectural Design of Airport Command Tower
Code 3: Code for Architectural Design of Electronic Information System Room
Code 4: Technical Specification for Airport Terminal Interior Decoration

工作室的蔓设计
The Organically-Permeated Design by WZJ Studio

2008.8.10 沙特馆团队正在制作工作模型

2008.9.12 沙特馆设计团队在沙特利雅得城乡事务部讨论方案

2008.9.15 王振军参加沙特馆执行官 Dr.Alghamdi 的招待晚宴

2009.3.9 王振军陪同沙特馆执行官 Dr.Alghamdi 视察工地

2010.5.14 王振军接受CCTV《世博的记忆》纪录片专访

2010.5.24 王振军与沙特馆项目管理方英国 ISG 公司德国建筑师 Cindy Burkhardt 合影

2010.5.25 王振军陪同中国工程院院士何镜堂、《建筑学报》副主编范雪参观沙特馆

2010.6.29 参加世博会最佳实践暨国际论坛，发表主题演讲

2010.9.5 王振军陪同中国建筑学会宋春华理事长参观沙特馆，并与馆长哈桑会谈

2010.9.30 沙特馆外排队观众

与顾问总建筑师张建元研讨方案

2017.6 初谈蔓设计

潍坊餐饮与展示中心三维曲面定位研究——三人行必有我师

创意总监李达带领团队讨论方案

副设计总监崔伟在审校北京新机场行政楼施工图——乐在其中

北京新机场行政与配套项目投标阶段——凌晨 12 点讨论方案

最后的封标时刻

工作室赴外地投标——带着"孩子"出征

郑州云计算项目投标阶段——冲刺时刻的 KFC

愉快的日常工间操

海盐工地考察——建筑完成前的再思考

项目三方在工地切磋

想看按摩医院场地却没开门

运营总监孙成伟在海盐工地

与材料商在工地探讨材料做法

设计总监陈珑在信达工地——建筑师的幸福时刻

行政总监邓涛与设计总监陈珑在非洲援建项目工地

参观香山校区隈严吾新作

工作室调研建筑材料——以手为尺

工作室参观天津大学新图书馆

材料确认前,考察材料商

工作室承担中国电子工程设计院第六届全国建筑展布展工作

教师节快乐——建筑师的双重身份

陪同老师 Percival 参观鸟巢

美国建筑考察——融合在建筑中

庆功宴

中国电子工程设计院第六届尤尼克斯羽毛球赛获得团队第二名

开心工作与健康生活

工作室合家欢

中国电子工程设计院元旦活动

工作室简介

创办于 2009 年的王振军工作室，是以中国电子工程设计院集团总建筑师王振军命名的直属建筑专业工作室，在高科技园区、超大型城市综合体、智能建筑、博览建筑、机场建筑、医养建筑、高端居住建筑、度假酒店等领域积累了丰富经验。工作室将一如既往地秉承"蔓设计"理念和态度，为成为具有深度研发能力以及优质品牌效益的高水准设计团队做出不懈坚持和努力。

Introduction of WZJ Studio

Established in 2009, WZJ Studio is headed by Wang Zhenjun, Chief Architect of China Electronics Engineering Design Institute Group (CEEDI). It aims to give play to his more than thirty years of hands-on experience in design for hi-tech parks, mega urban complex, intelligent office building, exhibition building, airport building, medical and rehabilitation building, high-end residential buildings, resort hotels, etc. The "Organically-Permeated Design" will always be a brief and refined term for them to follow. The creative practice of the Studio in design of boutique buildings in the architectural sense helps CEEDI retain its architectural design at a relatively higher academic level while completing the heavy workload of design tasks.

工作室成员

主持建筑师： 王振军
顾问总建筑师： 张建元

成员（按入职时间排名）：

邓涛	孙成伟	李达	权薇	刘嘉嘉	董召英	陈珑
李晶	朱谞	高寒	赵翀玺	夏璐	惠添添	徐彤
张宇嘉	王沛	刘海傲	崔伟	方雪	申明	李徉贝
朱冉	时菲	郑秉东	鲍亦林	胥丽娜	张良钊	赵玮璐

Members of WZJ Studio

Director : Wang Zhenjun
Consultant : Zhang Jianyuan

Members:

Deng Tao	Sun Chengwei	Li Da	Quan Wei	Liu Jiajia	Dong Zhaoying	Chen Long
Li Jing	Zhu Xu	Gao Han	Zhao Chongxi	Xia Lu	Hui Tiantian	Xu Tong
Zhang Yujia	Wang Pei	Liu Haiao	Cui Wei	Fang Xue	Shen Ming	Li Yangbei
Zhu Ran	Shi Fei	Zheng Bingdong	Bao Yilin	Xu Lina	Zhang Liangzhao	Zhao Weilu

致谢

感谢总院领导和同仁，感谢王振军工作室全体成员及亲属的共同付出；感谢广大客户、兄弟单位、专家朋友及社会各界人士的大力支持；感谢杨超英摄影师对工作室作品的拍摄呈现和陈家林先生对本书提供的翻译协助；感谢每位读者的关注见证，使得我们能够更坚定地秉持建筑情怀砥砺前行，最大限度地将蔓设计理念延伸至项目全过程当中，努力去实现无愧于时代的建筑作品。

Acknowledgements

We are very grateful to the leaders and colleagues of CEEDI and to all the members of WZJ Studio and their relatives for their unity of devotion0. We are very appreciative of the strong support from our clients, associate organizations, experts and friends, as well as people across a broad spectrum. We also thank photographer Yang Chaoying for his shooting of WZJ Studio works and Mr. Chen Jialin for his translation of the book. Last but not least, we grateful to our acknowledgement goes to every reader, too for their attention and witness. Their help and support has enabled us to forge ahead more steadfastly with the dream of an architect in mind, to carry out the organically-permeated design concept throughout the entire process of a project, and to work hard for the objective of creating architectural works worthy of the times.